DATA ACQUISITION AND SIGNAL PROCESSING FOR SMART SENSORS

DATA ACQUISITION AND SIGNAL PROCESSING FOR SMART SENSORS

Nikolay V. Kirianaki and **Sergey Y. Yurish**
International Frequency Sensor Association, Lviv, Ukraine

Nestor O. Shpak
Institute of Computer Technologies, Lviv, Ukraine

Vadim P. Deynega
State University Lviv Polytechnic, Ukraine

JOHN WILEY & SONS, LTD

Other Wiley Editorial Offices

John Wiley & Sons, Inc., 605 Third Avenue,
New York, NY 10158-0012, USA

Wiley-VCH Verlag GmbH, Pappelallee 3,
D-69469 Weinheim, Germany

John Wiley & Sons Australia, Ltd, 33 Park Road, Milton,
Queensland 4064, Australia

John Wiley & Sons (Asia) Pte Ltd, 2 Clementi Loop #02-01,
Jin Xing Distripark, Singapore 129809

John Wiley & Sons (Canada) Ltd, 22 Worcester Road,
Rexdale, Ontario M9W 1L1, Canada

Library of Congress Cataloging-in-Publication Data

Data acquisition and signal processing for smart sensors / Nikolay V. Kirianaki ... [et al.].
       p. cm.
     Includes bibliographical references and index.
     ISBN 0-470-84317-9 (alk. paper)
       1. Detectors. 2. Microprocessors. 3. Signal processing. 4. Automatic data collection
     systems. I. Kirianaki, Nikolai Vladimirovich.

    TA165. D38 2001
    681'.2–dc21   2001046912

British Library Cataloguing in Publication Data

A catalogue record for this book is available from the British Library

ISBN 0 470 84317 9

Typeset in 10/12pt Times by Laserwords Private Limited, Chennai, India
This book is printed on acid-free paper responsibly manufactured from sustainable forestry,
in which at least two trees are planted for each one used for paper production.

CONTENTS

PREFACE

Smart sensors are of great interest in many fields of industry, control systems, biomedical applications, etc. Most books about sensor instrumentation focus on the classical approach to data acquisition, that is the information is in the amplitude of a voltage or a current signal. Only a few book chapters, articles and papers consider data acquisition from digital and quasi-digital sensors. Smart sensors and microsensors increasingly rely on resonant phenomena and variable oscillators, where the information is embedded not in the amplitude but in the frequency or time parameter of the output signal. As a rule, the majority of scientific publications dedicated to smart sensors reflect only the technological achievements of microelectronics. However, modern advanced microsensor technologies require novel advanced measuring techniques.

Because data acquisition and signal processing for smart sensors have not been adequately covered in the literature before, this book aims to fill a significant gap.

This book is based on 40 years of the authors' practical experience in the design and creation of sensor instrumentation as well as the development of novel methods and algorithms for frequency–time-domain measurement, conversion and signal processing. Digital and quasi-digital (frequency, period, duty-cycle, time interval and pulse number output) sensors are covered in this book.

Research results, described in this book, are relevant to the authors' international research in the frame of different R&D projects and International Frequency Sensor Association (IFSA) activity.

Who Should Read this Book?

This book is aimed at PhD students, engineers, scientists and researchers in both academia and industry. It is especially suited for professionals working in the field of measuring instruments and sensor instrumentation as well as anyone facing new challenges in measuring, and those involved in the design and creation of new digital smart physical or chemical sensors and sensor systems. It should also be useful for students wishing to gain an insight into this rapidly expanding area. Our goal is to provide the reader with enough background to understand the novel concepts, principles and systems associated with data acquisition, signal processing and measurement so that they can decide how to optimize their sensor systems in order to achieve the best technical performances at low cost.

How this Book is Organized

This book has been organized into 10 chapters.

Chapter 1, *Smart sensors for electrical and non-electrical, physical and chemical quantities: the tendencies and perspectives*, describes the main advantages of frequency–time-domain signals as informative parameters for smart sensors. The chapter gives an overview of industrial types of smart sensors and contains classifications of quasi-digital sensors. Digital and quasi-digital (frequency, period, duty-cycle, time interval and pulse number output) sensors are considered.

Chapter 2, *Converters for different variables to frequency–time parameters of electric signals*, deals with different voltage (current)-to-frequency and capacitance-to-period (or duty-cycle) converters. Operational principles, technical performances and metrological characteristics of these devices are discussed from a smart sensor point of view in order to produce further conversion in the quasi-digital domain instead of the analog domain. The open and loop (with impulse feedback) structures of such converters are considered. (Figures 2.11, 2.12, 2.13, 2.14, 2.15 and some of the text appearing in Chapter 2, section 2.1, are reproduced from *New Architectures of Integrated ADC*, PDS '96 Proceedings. Reproduced by permission of Maciej Nowinski.)

Chapter 3, *Data acquisition methods for multichannel sensor systems*, covers multichannel sensor systems with cyclical, accelerated and simultaneous sensor polling. Data acquisition methods with time-division and space-division channelling are described. The chapter contains information about how to calculate the time-polling cycle for a sensor and how to analyse the accuracy and speed of data acquisition. Main smart sensor architectures are considered from a data acquisition point of view. Data transmitting and error protection on the basis of quasi-ternary cyclic coding is also discussed.

Chapter 4, *Methods of frequency-to-code conversion for smart sensors*, discusses traditional methods for frequency (period)-to-code conversion, including direct, indirect, combined, interpolation, Fourier conversion-based counting techniques as well as methods for phase-shift-to-code conversion. Such metrological characteristics as quantization error, conversion frequency range and conversion speed as well as advantages and disadvantages for each of the methods are discussed and compared.

Chapter 5, *Advanced and self-adapting methods of frequency-to-code conversion*, discusses reciprocal, ratiometric, constant elapsed time (CET), M/T, single-buffered, double-buffered and DMA transfer advanced methods. Comparative and cost-effective analyses are given. Frequency ranges, quantization errors, time of measurement and other metrological performances as well as hardware and software requirements for realization from a smart sensor point of view are described. This chapter is very important because it also deals with the concepts, principles and nature of novel self-adapting methods of dependent count (MDC) and the method with non-redundant reference frequency. The chapter covers main metrological performances including accuracy, conversion time, frequency range as well as software and hardware for MDC realization. Advanced conversion methods for frequencies ratio, deviations and phase shifts are also described. Finally, some practical examples and modelling results are presented.

Chapter 6, *Signal processing for quasi-digital smart sensors*, deals with the main frequency signal manipulations including multiplication, division, addition, subtraction, derivation, integration and scaling. Particular attention has been paid to new methods of frequency multiplication and scaling with the aim of frequency signal unification. Different wave shapes (sine wave, sawtooth, triangular and rectangular) of a sensor's output are considered. It is also shown how the weight function averaging can be used for noise and quantization error reduction.

Chapter 7, *Digital output smart sensors with software-controlled performances and functional capabilities*, discusses program-oriented methods for frequency-, period-, duty-cycle-, time-interval-, phase-shift- and pulse-number-to-code conversion and digital smart sensors. The design methodology for optimal program-oriented conversion methods, correction of systematic errors and the modified method of algorithms merging are considered. Examples are given. This chapter also describes specific errors and features.

Chapter 8, *Multichannel intelligent and virtual sensor systems*, describes smart sensor systems with time- and space-division frequency channelling. Both are based on the method of dependent count. Comparative analysis is given. Performances and features are illustrated by an ABS smart sensor microsystem example. Multiparameters sensors are also considered. The chapter includes information about virtual sensor instrumentation and how to estimate the total error of arranged system. Definitions and examples (temperature, pressure, rotation speed virtual instruments) are given.

Chapter 9, *Smart Sensor Design at Software level*, deals with embedded microcontroller set instruction minimization for metering applications (to save chip area) and low-power design techniques—optimal low-power programming (for power consumption reduction). Many practical 'hints' (e.g. instruction selection and ordering, jump, call and cycle optimization, etc.), recommendations and examples are given.

Chapter 10, *Smart sensor buses and interface circuits*, describes sensor buses and network protocols from the smart sensor point of view. Modern sensor interface circuits are discussed. Particular attention has been given to the Universal Transducer Interface (UTI) and Time-to-Digital Converter (TDC), which allow low-cost interfacing with different analog sensors elements such as Pt resistors, thermistors, potentiometer resistors, capacitors, resistive bridges, etc. and convert analog sensor signals to the quasi-digital domain (duty-cycle or time interval).

Finally, we discuss what the future might bring.

References. Apart from books, articles and papers, this section includes a large collection of appropriate Internet links, collected from the Sensors Web Portal launched by the authors.

LIST OF ABBREVIATIONS AND SYMBOLS

δ_q	program-specified relative quantization error
Δ_q	absolute quantization error
D_f	specified measuring range of frequencies
f_x	measurand frequency
f_0	reference frequency
F	greater of the two frequencies f_x and f_0
f	lower of the two frequencies f_x and f_0
f_{bound}	lower frequency limit
F_{bound}	upper frequency limit
m	counter capacity
N_δ	number, determined by the error $\delta = 1/N_\delta$
N_x	number of periods of lower frequency f
T	period of greater frequency ($T = 1/F$)
τ	period of lower frequency ($\tau = 1/f$)
T_q	quantization window
T_0	reference gate time interval
ABS	antilock braking system
ADC	analog-to-digital converter
ALU	arithmetic logic unit
ASIC	application specific integrated circuit
ASIP	application specific instruction processor
CAD	computer-aided design
CMOS	complementary metal oxide semiconductor
CT	counter
DAC	digital-to-analog converter
DAQ	data acquisition
DFT	discrete Fourier transformation
DSP	digital signal processor
FCC	frequency-to-code converter
FPGA	field-programmable gate array
FS	full scale
GUI	graphical user interface
LCF	Liapunov characteristic function

MDC	method of dependent count
μK	microcontroller
μP	microprocessor
MSM	multichip module
PCA	programable counter array
PCM	program-oriented conversion method
PWM	pulse width modulation
RAM	random access memory
ROM	read-only memory
VFC	voltage-to-frequency converter
VLSI	very large scale integration

INTRODUCTION

Rapid advances in IC technologies have brought new challenges to the physical design of integrated sensors and micro-electrical-mechanical systems (MEMS). Microsystem technology (MST) offers new ways of combining sensing, signal processing and actuation on a microscopic scale and allows both traditional and new sensors to be realized for a wide range of applications and operational environments. The term 'MEMS' is used in different ways: for some, it is equivalent to 'MST', for others, it comprises only surface-micromechanical products. MEMS in the latter sense are seen as an extension to IC technology: 'an IC chip that provides sensing and/or actuation functions in addition to the electronic ones' [1]. The latter definition is used in this book.

The definition of a smart sensor is based on [2] and can be formulated as: 'a smart sensor is one chip, without external components, including the sensing, interfacing, signal processing and intelligence (self-testing, self-identification or self-adaptation) functions'.

The main task of designing measuring instruments, sensors and transducers has always been to reach high metrology performances. At different stages of measurement technology development, this task was solved in different ways. There were technological methods, consisting of technology perfection, as well as structural and structural-algorithmic methods. Historically, technological methods have received prevalence in the USA, Japan and Western Europe. The structural and structural-algorithmic methods have received a broad development in the former USSR and continue developing in NIS countries. The improvement of metrology performances and extension of functional capabilities are being achieved through the implementation of particular structures designed in most cases in heuristic ways using advanced calculations and signal processing. Digital and quasi-digital smart sensors and transducers are not the exception.

During measurement different kinds of measurands are converted into a limited number of output parameters. Mechanical displacement was the first historical type of such (unified) parameters. The mercury thermometer, metal pressure gauge, pointer voltmeter, etc. are based on such principles [3]. The amplitude of an electric current or voltage is another type of unified parameter. Today almost all properties of substances and energy can be converted into current or voltage with the help of different sensors. All these sensors are based on the use of an amplitude modulation of electromagnetic processes. They are so-called analog sensors.

Digital sensors appeared from a necessity to input results of measurement into a computer. First, the design task of such sensors was solved by transforming an analog quantity into a digital code by an analog-to-digital converter (ADC). The creation of quasi-digital sensors, in particular, frequency sensors, was another very promising direction [3]. *Quasi-digital sensors are discrete frequency–time-domain sensors with frequency, period, duty-cycle, time interval, pulse number or phase shift output.* Today, the group of frequency output sensors is the most numerous among all quasi-digital sensors (Figure I). Such sensors combine a simplicity and universatility that is inherent to analog devices, with accuracy and noise immunity, proper to sensors with digital output. Further transformation of a frequency-modulated signal was reduced by counting periods of a signal during a reference time interval (gate). This operation exceeds all other methods of analog-to-digital conversion in its simplicity and accuracy [4].

Separate types of frequency transducers, for example, string tensometers or induction tachometers, have been known for many years. For example, patents for the string distant thermometer (Patent No. 617 27, USSR, Davydenkov and Yakutovich) and the string distant tensometer (Patent No. 21 525, USSR, Golovachov, Davydenkov and Yakutovich) were obtained in 1930 and 1931, respectively. However, the output frequency of such sensors (before digital frequency counters appeared) was measured by analog methods and consequently substantial benefit from the use of frequency output sensors was not achieved practically.

The situation has changed dramatically since digital frequency counters and frequency output sensors attracted increasing attention. As far back as 1961 Professor P.V. Novitskiy wrote: '... In the future we can expect, that a class of frequency sensors will get such development, that the number of now known frequency sensors will exceed the number of now known amplitude sensors...' [3]. Although frequency output sensors exist practically for any variables, this prognosis has not yet been fully justified for various reasons.

With the appearance in the last few years of sensor microsystems and the heady development of microsystem technologies all over the world, technological and cost factors have increased the benefits of digital and quasi-digital sensors. Modern technologies are able to solve rather complicated tasks, concerned with the creation of different sensors. Up to now, however, there have still been some major obstacles preventing industries from largely exploiting such sensors in their systems. These are only some subjective reasons:

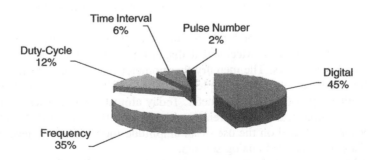

Figure I Classification of sensors from discrete group in terms of output signals (IFSA, 2001)

- The lack of awareness of the innovation potential of modern methods for frequency-time conversion in many companies, as processing techniques have mainly been developed in the former Soviet Union.

- The tendency of companies to return, first of all, major expenditures, invested in the development of conventional ADCs.

- The lack of emphasis placed on business and market benefits, which such measuring technologies can bring to companies etc.

Today the situation has changed dramatically. According to Intechno Consulting, the non-military world market for sensors has exceeded expectations with US$32.5 billion in 1998. By 2003, this market is estimated to grow at an annual rate of 5.3% to reach US$42.2 billion. Under very conservative assumptions it is expected to reach US$50–51 billion by 2008; assuming more favourable but still realistic economic conditions, the global sensor market volume could even reach US$54 billion by 2008. Sensors on a semiconductor basis will increase their market share from 38.9% in 1998 to 43% in 2008. Strong growth is expected for sensors based on MEMS-technologies, smart sensors and sensors with bus capabilities [5]. It is reasonable to expect that silicon sensors will go on to conquer other markets, such as the appliances, telecommunications and PC markets [6].

We hope that this book will be a useful and relevant resource for anyone involved in the design of high performance and highly efficient digital smart sensors and data acquisition systems.

1

SMART SENSORS FOR ELECTRICAL AND NON-ELECTRICAL, PHYSICAL AND CHEMICAL VARIABLES: TENDENCIES AND PERSPECTIVES

The processing and interpretation of information arriving from the outside are the main tasks of data acquisition systems and measuring instruments based on computers. Data acquisition and control systems need to get real-world signals into the computer. These signals come from a diverse range of transducers and sensors. According to [7] 'Data Acquisition (DAQ) is collecting and measuring electrical signals from sensors and transducers and inputting them to a computer for processing.' Further processing can include the sensors' characteristic transformation, joint processing for many parameters as well as statistical calculation of results and presenting them in a user-friendly manner.

According to the output signal, sensors and transducers can be divided into potential (amplitude), current, frequency, pulse-time and code. As a result, the task of adequate sensor interfacing with PCs arises before the developers and users of any data acquisition systems. Therefore special attention must be paid to the problems of output conversion into a digital format as well as to high accuracy and speed conversion methods.

In general, a sensor is a device, which is designed to acquire information from an object and transform it into an electrical signal. A classical integrated sensor can be divided into four parts as shown in Figure 1.1. The first block is a sensing element (for example, resistor, capacitor, transistor, piezo-electric material, photodiode, resistive bridge, etc.). The signal produced from the sensing element itself is often influenced by noise or interference. Therefore, signal-conditioning and signal-processing techniques such as amplification, linearization, compensation and filtering are necessary (second block) to reduce sensor non-idealities.

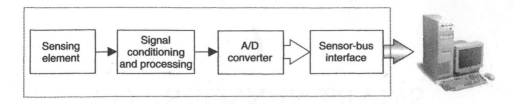

Figure 1.1   Classical integrated sensors

Sometimes if certain sensing elements are used on the same chip, a multiplexer is necessary. In cases of data acquisition, the signal from the sensor must be in a serial or parallel digital format. This function can be realized by the analog-to-digital or frequency-to-digital converter. The last (but not least) block is a sensor-bus interface. A data acquisition system can have a star configuration in which each sensor is connected to a digital multiplexer. When using a large number of sensors, the total cable length and the number of connections at the multiplexer can become very high. For this reason it is much more acceptable to have a bus-organized system, which connects all data sources and receivers. This bus system handles all data transports and is connected to a suitable interface that sends accumulated data to the computer [8].

A smart sensor block diagram is shown in Figure 1.2. A microcontroller is typically used for digital signal processing (for example, digital filtering), analog-to-digital or frequency-to-code conversions, calculations and interfacing functions. Microcontrollers can be combined or equipped with standard interface circuits. Many microcontrollers include the two-wire I^2C bus interface, which is suited for communication over short distances (several metres) [9] or the serial interface RS-232/485 for communication over relatively long distances.

However, the essential difference of the smart sensor from the integrated sensor with embedded data-processing circuitry is its intelligence capabilities (self-diagnostics, self-identification or self-adaptation (decision-making)) functions. As a rule, these functions are implemented due to a built-in microcontroller (microcontroller core ('microcontroller like' ASIC) or application-specific instruction processor (ASIP)) or DSP. New functions and the potential to modify its performance are the main advantages of smart sensors. Due to smart sensor adaptability the measuring process can be optimized for maximum accuracy, speed and power consumption. Sometimes 'smart sensors' are called 'intelligent transducers'.

At present, many different types of sensors are available. Rapid advancement of the standard process for VLSI design, silicon micromachining and fabrication provide the technological basis for the realization of smart sensors, and opens an avenue that

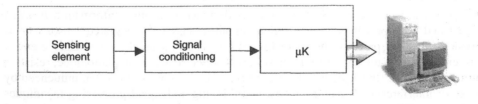

Figure 1.2   Smart sensor

can lead to custom-integrated sensors to meet the new demands in performance, size and cost. This suggests a smooth merging of the sensor and electronics and the fabrication of complete data acquisition systems on a single silicon chip. The essential issue is the fabrication compatibility of the sensor, sensor-related analogue microelectronic circuits and digital interface circuits [10]. In fact, for any type of silicon sensing element and read-out circuitry, a process can be developed to merge them onto a single chip. However, process development is very expensive and therefore only a huge production volume will pay off the development cost. Successful integrated-sensor processes must have an acceptable complexity and/or applicability for a wide range of sensors [11]. MEMS technologies allow the miniaturization of sensors and, at the same time, integration of sensor elements with microelectronic functions in minimal space. Only MEMS technologies make it possible to mass-produce sensors with increasing cost-effectiveness while improving their functionality and miniaturizing them.

Of course, the implementation of the microcontroller in one chip together with the sensing element and signal-conditioning circuitry is an elegant and preferable engineering solution. However, the combination of monolithic and hybrid integration with advanced processing and conversional methods in many cases achieves magnificent technical and metrological performances for the shorter time-to-market period without additional expenditures for expensive CAD tools and the lengthy smart sensor design process. For implementation of smart sensors with hybrid-integrated processing electronics, hardware minimization is a necessary condition to achieve a reasonable price and high reliability. In this case, we have the so-called 'hybrid smart sensor' in which a sensing element and an electronic circuit are placed in the same housing.

Frequency–time-domain sensors are interesting from a technological and fabrication compatibility point of view: the simplifications of the signal-conditioning circuitry and measurand-to-code converter, as well as metrology performances and the hardware for realization. The latter essentially influences the chip area. Such sensors are based on resonant phenomena and variable oscillators, whose information is embedded not in the amplitude but in the frequency or the time parameter of the output signal. These sensors have frequency (f_x), period ($T_x = 1/f_x$), pulse width (t_p), spacing interval (t_s), the duty cycle (t_p/T_x), online time ratio or off-duty factor (T_x/t_p), pulse number (N), phase shift (φ) or single-time interval (τ) outputs. These informative parameters are shown in Figure 1.3. Because these informative parameters have analog and digital signal properties simultaneously, these sensors have been called 'quasi-digital'. Frequency output sensors are the most numerous group among all quasi-digital sensors (see Figure I, Introduction). Let us consider the main advantages of frequency as the sensor's output signal.

- *High noise immunity*. In frequency sensors, it is possible to reach a higher accuracy in comparison with analog sensors with analog-to-code conversion. This property of high noise immunity proper to a frequency modulation is apparently the principal premise of frequency sensors in comparison with analog ones. The frequency signal can be transmitted by communication lines for a much greater distance than analog and digital signals. The frequency signal transmitted practically represents a serial digital signal. Thus, all the advantages of digital systems are demonstrated. Also, only a two-wire line is necessary for transmission of such a signal. In comparison with the usual serial digital data transmission it has the advantage of not

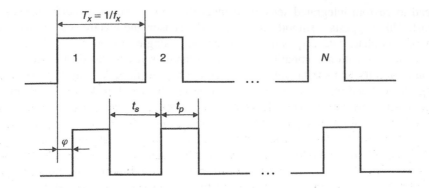

Figure 1.3   Informative parameters of frequency–time-domain sensors

requiring any synchronization. A frequency signal is ideal for high noise industrial environments.

- *High output signal power.* The sensor's signal can be grouped into six energy domains: electrical, thermal, mechanical, chemical, radiant and magnetic. Electrical signals are currently the most preferred signal form. Therefore, sensor design is focused on developing transducers that convert the signal from one or other energy domain into a quantity in the electrical domain. From the power point of view, the section from a sensor output up to an amplifier input is the heaviest section in a measuring channel for transmitting signals. Here the signal is transmitted by a very small level of energy. The losses, originating in this section, cannot be filled any more by signal processing. Output powers of frequency sensors are, as a rule, considerably higher. In this case, the power affecting the generated frequency stability is the oscillation (reactive) power of the oscillating loop circuit and due to the higher quality factor of the oscillating loop its power is higher.

- *Wide dynamic range.* Because the signal is in the form of the frequency, the dynamic range is not limited by the supply voltage and noise. A dynamic range of over 100 dB may be easily obtained.

- *High accuracy of frequency standards.* The frequency reference, for example, crystal oscillators, can be made more stable than the voltage reference. This can be explained in the same way as information properties of amplitude-modulated and frequency-modulated signals.

- *Simplicity of commutation and interfacing.* Parasitic emf, transient resistances and cross-feed of channels in analog multiplexers by analog sensors are reduced to the occurrence of complementary errors. The frequency-modulated signal is not sensitive to all the above listed factors. Multiplexers for frequency sensors and transducers are simple enough and do not introduce any errors into observed results.

- *Simplicity of integration and coding.* The precise integration in time of frequency sensors' output signals can be realized simply enough. The adding pulse counter is an ideal integrator with an unlimited measurement time. The frequency signal can be processed by microcontrollers without any additional interface circuitry.

All this makes the design and usage of different frequency–time-domain smart sensors very efficient.

The most important properties of smart sensors have been well described in [9]. Here we will briefly describe only the basic focus points for an intelligent frequency–time-domain smart sensor design.

- *Adaptability.* A smart sensor should be adaptive in order to optimize the measuring process. For example, depending on the measuring conditions, it is preferable to exchange measurement accuracy for speed and conversely, and also to moderate power consumption, when high speed and accuracy are not required. It is also desirable to adjust a clock crystal oscillator frequency depending on the environment temperature. The latter also essentially influences the following focus point–the accuracy.

- *Accuracy.* The measuring error should be programmable. Self-calibration will allow reduction of systematic error, caused, for example, by the inaccuracy of the system parameters. The use of statistical algorithms and composited algorithms of the weight average would allow reduction of random errors caused, for instance, by interference, noise and instability.

- *Reliability.* This is one of the most important requirements especially in industrial applications. Self-diagnostics is used to check the performance of the system and the connection of the sensor wires.

For analysis of quasi-digital smart sensors, it is expedient to use the classifica...on, shown in Figure 1.4. Depending on conversion of the primary information into frequency, all sensors are divided into three groups: sensors with measurand-to-frequency conversion; with measurand-to-voltage-to-frequency conversion; and with measurand-to-parameter-to-frequency conversion.

1. *Sensors with $x(t) \rightarrow F(t)$ conversion.* These are sensors that generate a frequency output. Electronic circuitry might be needed for the amplification of the impedance matching, but it is not needed for the frequency conversion step itself. Measuring information like the frequency or the frequency-pulse form is most simply obtained in inductive, photoimpulse, string, acoustic and scintillation sensors, since the principle of operation allows the direct conversion $x(t) \rightarrow F(t)$. One group of

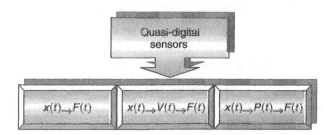

Figure 1.4 Quasi-digital smart sensor classification ($x(t)$ — measurand; $F(t)$ — frequency; $V(t)$ — voltage, proportional to the measurand; $P(t)$ — parameter)

such sensors is based on resonant structures (piezoelectric quartz resonators, SAW (surface acoustic wave) dual-line oscillators, etc.) whereas another group is based on the periodic geometrical structure of the sensors, for example, angle encoders.

2. *Sensors with $x(t) \rightarrow V(t) \rightarrow F(t)$ conversion.* This group has a numerous number of different electric circuits. These are Hall sensors, thermocouple sensors and photosensors based on valve photoelectric cells. When a frequency output is required, a simple voltage-to-frequency or current-to-frequency conversion circuit can be applied to obtain the desired result.

3. *Sensors with $x(t) \rightarrow P(t) \rightarrow F(t)$ conversion.* The sensors of this group are rather manifold and numerous. These are the so-called electronic-oscillator based sensors. Such sensors are based on the use of electronic oscillators in which the sensor element itself is the frequency-determining element. These are inductive, capacity and ohmic parametric (modulating) sensors.

Parametric (modulating) sensors are devices that produce the primary information by way of respective alterations of any electrical parameter of some electrical circuit (inductance, capacity, resistance, etc.), for which it is necessary to have an external auxiliary power supply. Examples of such types of sensors are pressure sensors based on the piezoresistive effect and photodetectors based on the photoelectric effect.

In turn, *self-generating sensors* are devices that receive a signal immediately by way of a current $i(t)$ or voltage $V(t)$ and do not require any source of power other than the signal being measured. Examples of such types of sensors are Seebeck effect based thermocouples and photoeffect based solar cells. Self-generating sensors are also called 'active' sensors, while modulating sensors are called 'passive' sensors.

The signal power of modulating sensors is the largest and, therefore, from the noise-reduction point of view their usage is recommended.

The distinctiveness of these three sensor groups (Figure 1.4) is the absence of conventional ADCs. In order to design digital output smart sensors, it is expedient to use a microcontroller for the frequency-to-code conversion. The production of such smart sensors does not require extra technological steps. Moreover, modern CAD tools contain microcontroller cores and peripheral devices as well as voltage-to-frequency converters (VFC) in the library of standard cells. So, for example, the Mentor Graphic CAD tool includes different kinds of VFCs like AD537/650/652, the CAD tool from Protel includes many library cells of different Burr–Brown VFCs.

In comparison with the data-capturing method using traditional analog-to-digital converters (ADC), the data-capturing method using VFC has the following advantages [12]:

- Simple, low-cost alternative to the A/D conversion.

- Integrating input properties, excellent accuracy and low nonlinearity provide performance attributes unattainable with other converter types, make VFC ideal for high noise industrial environments.

- Like a dual-slope ADC, the VFC possesses a true integrating input and features the best, much better than a dual-slope converter, noise immunity. It is especially important in industrial measurement and data acquisition systems. While a successive approximation ADC takes a 'snapshot' in time, making it susceptible to noise peaks, the VFC's input is constantly integrating, smoothing the effects of noise or varying input signals.

- It has universality. First, its user-selected voltage input range ($\pm$ supply). Second, the high accuracy of the frequency-to-code conversion (up to 0.001%). The error of such conversion can be neglected in a measuring channel. This is not true for traditional analog-to-digital conversion. The ADC error is commensurable with the sensor's error, especially if we use modern high precision sensors with relative error up to 0.01%.

- When the data-capturing method with a VFC is used, a frequency measurement technique must also be chosen which meets the conversion speed requirements. While it is clearly not a 'fast' converter in a common case, the conversion speed of a VFC system can be optimized by using efficient techniques. Such optimization can be performed due to advanced methods of frequency-to-code conversion, quasi pipeline data processing in a microcontroller and the use of novel architectures of VFC.

While pointing out the well-known advantages of frequency sensors, it is necessary to note, that the number of physical phenomena, on the basis of which sensors with frequency and digital outputs can be designed, is essentially limited. Therefore, analog sensors with current and voltage outputs have received broad dissemination. On the one hand, this is because of the high technology working off analog sensor units, and also because of the heady development of analog-to-digital conversion in the last few years. On the other hand, voltage and current are widely used as unified standard signals in many measuring and control systems.

An important role is played by the technological and cost factors in the choice of sensor. Therefore, the question, what sensors are the best–frequency or analog–is not enough. With the appearance of sensor microsystems and the heady development of microsystem technologies all over the world, technological and cost factors need to be modified for the benefit of frequency sensors.

Sensor types with the highest demand volumes are temperature sensors, pressure sensors, flow sensors, binary position sensors (proximity switches, light barriers, reflector-type photosensors), position sensors, chemical sensors for measurement in liquids and gases, filling sensors, speed and rpm sensors, flue gas sensors and fire detectors worldwide. The fastest growing types of sensors include rain sensors, thickness sensors, sensors that measure the quality of liquids, navigation sensors, tilt sensors, photodetectors, glass breakage sensors, biosensors, magnetic field sensors and motion detectors [5].

The frequency–time-domain sensor group is constantly increasing. First, it is connected with the fast development of modern microelectronic technologies, secondly with the further development of methods of measurement for frequency-domain parameters of signals and methods for frequency-to-code conversion, and thirdly, with advantages of frequency as the informative parameter of sensors and transducers. Today it is difficult to find physical or chemical variables, for which frequency output or digital sensors do not exist. Of course, this book cannot completely describe all existing

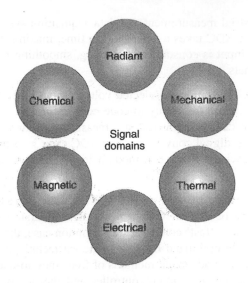

Figure 1.5 Six signal domains

sensors and their principles of operation. For readers wishing to learn more about smart sensor development history we would like to recommend the article [6].

This review aims to illustrate state-of-the-art frequency–time-domain IC sensors with high metrology performances, and also to formulate the basic requirements for such an important smart sensor's unit, as the frequency-to-code converter.

Frequency–time domain-sensors can be grouped in several different ways. We will group them according to the measurand domains of the desired information. There are six signal domains with the most important physical parameters shown in Figure 1.5. Electrical parameters usually represent a signal from one of the non-electrical signal domains.

1.1 Temperature IC and Smart Sensors

Temperature sensors play an important role in many measurements and other integrated microsystems, for example, for biomedical applications or self-checking systems and the design for the thermal testability (DfTT). IC temperature sensors take advantage of the variable resistance properties of semiconductor materials. They provide good linear frequency, the duty-cycle or pulse width output proportional to the temperature typically in the range from $-55\,°C$ to $+150\,°C$ at a low cost. These devices can provide direct temperature readings in a digital form, thus eliminating the need for an ADC. Because IC sensors can have a memory, they can be very accurately calibrated, and may operate in multisensor environments in applications such as communications networks. Many IC sensors also offer communication protocols for use with bus-type data acquisition systems; some also have addressability and data storage and retrieval capabilities.

Smart temperature sensors need to be provided with some kind of output digital signal adapted to microprocessors and digital-processing systems. This signal can be a time-signal type, where the measurement is represented by the duty-cycle or the frequency ratio, or the fully digital code that is sent to the processor in a serial way through the digital bus [13]. Some important restraints, caused by the integration of sensing and digital-processing function on the same chip are [14] (a) the limited chip area, (b) the tolerances of the device parameters and (c) the digital interference. Performances of some integrated temperature sensors are shown in Table 1.1.

Since CMOS is still the most extensively used technology the integration of temperature sensors in high-performance, low-cost digital CMOS technologies is preferred in order to allow signal conditioning and digital processing on the same chip [13].

In the framework of the COPERNICUS EC project CP0922, 1995–1998, THER-MINIC (THermal INvestigations of ICs and Microstructures), the research group from Technical University of Budapest has dealt for several years with the design problem of small-size temperature sensors that must be built into the chip for thermal monitoring [18–21]. One such sensor is based on a current-to-frequency converter [18,19]. The analog signal of the current output CMOS sensor is converted into a quasi-digital one using a current-to-frequency converter. The block diagram of the frequency output sensor is shown in Figure 1.6. The I_{out} output current and its 'copy' generated by a current mirror the charge and discharge the capacitor C_x. The signal of the capacitor is led to a differential comparator the reference voltage of which is switched between the levels V_C and V_D [19]. The resulting frequency is

$$f = \frac{I_{out}}{2 \cdot C_x (V_C - C_D)}. \tag{1.1}$$

The sensitivity is $-0.808\%/°C$. The output frequency 0.5–1.3 MHz is in a convenient range. The complete circuit requires only an area of 0.018 mm^2 using the ECPD 1 μm CMOS process. The low sensitivity on the supply voltage is a remarkable feature: ± 0.25 V charge in V_{DD} results in only $\pm 0.28\%$ charge in the frequency. The latter corresponds to the $\pm 0.35\,°C$ error. The total power consumption of this sensor is about 200 μW [20].

The characteristic of this sensor is quite linear and the output frequency of these sensors can be approximately written as

$$f_{out} = f_{20Cels} \exp(\gamma (T_{Cels} - 20\,°C)), \tag{1.2}$$

Table 1.1 Performances of some integrated temperature sensors

Sensor	Output type	Characteristic	Area, mm^2	IC technology
[15]	Digital	I → F converter + DSP	4.5	CMOS
[16]	Duty-cycle	Duty-cycle-modulated	5.16	Bipolar
[17]	Frequency	I → F converter	6	Bipolar

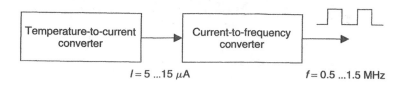

Figure 1.6 Block diagram of the temperature frequency output sensor

in the $-50\cdots+120\,°C$ temperature range, where γ is the sensitivity, f_{20Cels} is the nominal frequency related to $T = 20\,°C$. Using the AMS 0.8 µm process, the area consumption is 0.005 mm^2 [21]. The THSENS-F [22] sensor characteristic and sensor layout based on these researches are shown in Figure 1.7 and 1.8 respectively. This sensor can be inserted into CMOS designs, which can be transferred and re-used as cell (layout level) entities or as circuit netlists with transistor sizes. The sensor's sensitivity is $\approx -0.8\%/°C$; the temperature range is $-50\cdots+150\,°C$; the accuracy is $\approx \pm 2\,°C$ for $(0\ldots120\,°C)$. The two latter parameters depend on the process.

If the temperature sensors described above are inserted into a chip design, additional circuitry must be implemented in order to provide access to such sensors [21]. Built-in temperature sensors can be combined with other built-in test circuitry. The boundary scan architecture [23] is suitable for monitoring temperature sensors. This architecture has led to the standard IEEE 1149.1 and is suitable for incorporating frequency output temperature sensors.

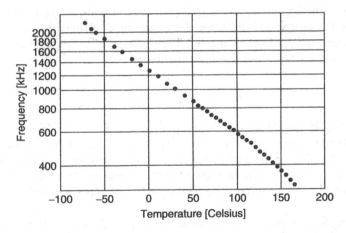

Figure 1.7   Sensor characteristic (output frequency vs. temperature) (Reproduced by permission of MicReD)

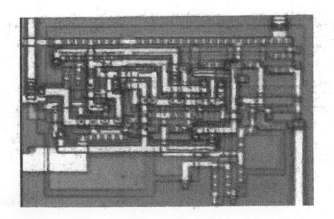

Figure 1.8   Sensor layout (Reproduced by permission of MicReD)

A further interesting fully CMOS temperature sensor designed by this research team is based on the temperature dependence of internal thermal diffusion constant of silicon. In order to measure this diffusion constant an oscillating circuit is used in which the frequency-determining element is realized by a thermal delay line. The temperature difference sensors used in this delay line are Si-Al thermopiles. This circuit is the Thermal-Feedback Oscillator (TFO). The frequency of this oscillator is directly related to the thermal diffusion constant and thus to the temperature. This constant can be defined as

$$D_{th} = \lambda/c, \tag{1.3}$$

where λ is the thermal conductivity and c is the unit-volume heat capacitance. This diffusion constant shows a reasonably large $(-0.57\%/°C)$ temperature dependence on the silicon. In order to measure this diffusion constant, oscillating circuits were used in which the frequency-determining feedback element is realized by a thermal time-delay line. If the feedback element is a thermal two-port (thermal delay line) then the frequency of the oscillator is directly related to the thermal diffusion constant and thus shows similar temperature dependence as the thermal diffusion constant [18].

The thermal delay line requires, however, a significant power input. Because of this disadvantage, the circuit is not really suitable for online monitoring purposes [19]. However, these sensors and the sensor principle can probably be used for other applications.

It is also necessary to note the low-power consumption smart temperature sensor SMT 160-30 from Smartec (Holland) [24]. It is a sophisticated full silicon sensor with a duty-cycle modulated square-wave output. The duty-cycle of the output signal is linearly related to the temperature according to the equation:

$$DC = \frac{t_p}{T_x} = t_p \cdot f_x = 0.320 + 0.00470 \cdot t, \tag{1.4}$$

where t_p is the pulse duration; T_x is the period; f_x is the frequency; t is the temperature in °C. This sensor is calibrated during the test and burn-in of the chip. The sensor characteristic is shown in Figure 1.9.

One wire output can be directly connected to all kinds of microcontrollers without the A/D conversion. The temperature range is $-45°C$–$+150°C$, the best absolute

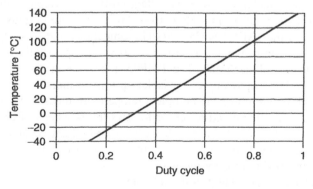

Figure 1.9 Sensor characteristic (temperature vs. duty-cycle) (Reproduced by permission of Smartec)

accuracy, including all errors is $\pm 0.7\,°C$, the relative error is 0.47%, the frequency range is 1–4 kHz. The CMOS output of the sensor can handle cable length up to 20 metres. This makes the SMT 160-30 very useful for remote sensing and control applications. This smart temperature sensor represents a significant totally new development in transducer technology. Its novel on-chip interface meets the progressively stringent demands of both the consumer and industrial electronics sectors for a temperature sensor directly connectable to the microprocessor input and thus capable of direct and reliable communication with microprocessors.

In applications where more sensors are used, easy multiplexing can be obtained by using more microprocessor inputs or by using simple and cheap digital multiplexers.

The next specialized temperature sensor is interesting due to the high metrology performances (high accuracy). It is the SBE 3F temperature sensor with an initial accuracy of $0.001\,°C$ and typically stable to $0.002\,°C$ per year [25]. It is used for custom-built oceanographic profiling systems or for high-accuracy industrial and environmental temperature-monitoring applications. Depth ratings to 6800 and 10 500 metres (22 300 and 34 400 ft) are offered to suit different application requirements.

The sensing element is a glass-coated thermistor bead, pressure-protected in a thin-walled 0.8 mm diameter stainless steel tube. Exponentially related to the temperature, the thermistor resistance is the controlling element in the optimized Wien bridge oscillator circuit. The resulting sensor frequency is inversely proportional to the square root of the thermistor resistance and ranges from approximately 2 to 6 kHz, corresponding to temperatures from -5 to $+35\,°C$.

In speaking about digital output IC and smart temperature sensors it is necessary to mention interesting developments of companies such as Analog Devices, Dallas Semiconductor and National Semiconductor.

The TMP03/TMP04 are monolithic temperature detectors from Analog Devices [26,27] that generate a modulated serial digital output that varies in direct proportion to the temperature of the device. The onboard sensor generates a voltage precisely proportional to the absolute temperature, which is compared with the internal voltage reference and the input to a precision digital modulator. The ratiometric encoding format of the serial digital output is independent from the clock drift errors common to most serial modulation techniques such as voltage-to-frequency converters. The overall accuracy is $\pm 1.5\,°C$ (typical) from $-25\,°C$ to $+100\,°C$, with good transducer linearity. The digital output of the TMP04 is CMOS/TTL compatible, and is easily interfaced to the serial inputs of the most popular microprocessors. The open-collector output of the TMP03 is capable of sinking 5 mA. The TMP03 is best suited for systems requiring isolated circuits, utilizing optocouplers or isolation transformers.

The TMP03/TMP04 are powerful, complete temperature measurement systems on a single chip. The onboard temperature sensor follows the footsteps of the TMP01 low-power programmable temperature controller, offering excellent accuracy and linearity over the entire rated temperature range without correction or calibration by the user.

The sensor output is digitized by the first-order sigma-delta modulator, also known as the 'charge balance' type analog-to-digital converter. This type of converter utilizes the time-domain oversampling and a high accuracy comparator to deliver 12 bits of effective accuracy in the extremely compact circuit.

Basically, the sigma-delta modulator consists of an input sampler, a summing network, an integrator, a comparator, and a 1-bit DAC. Similar to the voltage-to-frequency converter, this architecture creates in effect a negative feedback loop whose intent is to minimize the integrator output by changing the duty-cycle of the comparator output in response to input voltage changes. The comparator samples the output of the integrator at a much higher rate than the input sampling frequency, called the oversampling. This spreads the quantization noise over a much wider band than that of the input signal, improving the overall noise performance and increasing the accuracy.

The modulated output of the comparator is encoded using a circuit technique (patent pending), which results in a serial digital signal with a mark-space ratio format that is easily decoded by any microprocessor into either degrees centigrade or degrees Fahrenheit values, and is readily transmitted or modulated over a single wire. It is very important that this encoding method avoids major error sources common to other modulation techniques, as it is clock-independent.

The AD7818 (single-channel) and AD7817 (4-channel) [28,29] are on-chip temperature sensors with 10-bit, single and four-channel A/D converters. These devices contain an 8 ms successive-approximation converter based around a capacitor DAC, an on-chip temperature sensor with an accuracy of $\pm1\,°C$, an on-chip clock oscillator, inherent track-and-hold functionality and an on-chip reference (2.5 V $\pm\,0.1\%$). Some other digital temperature sensors from Analog Devices [30] are shown in Table 1.2.

Dallas Semiconductor offers a broad range of factory-calibrated 1-, 2-, 3- Wire® or SPI buses temperature sensors/thermometers that can provide straightforward thermal management for a vast array of applications. This unparalleled product line includes a variety of 'direct-to-digital' temperature sensors that have the accuracy and features to easily improve system performance and reliability [31]. These devices reduce the component count and the board complexity by conveniently providing digital data without the need for dedicated ADCs. These sensors are available with accuracies ranging from $\pm0.5\,°C$ to $\pm2.5\,°C$ (guaranteed over wide temperature and power-supply ranges), and they can operate over a temperature range of $-50\,°C$ to $+125\,°C$.

The conversion time range for the temperature into a digital signal is 750 ms-1.2 s. The 1-Wire and 2-Wire devices have a multi-drop capability, which allows multiple sensors to be addressed on the same bus. In addition, some devices (DS1624, DS1629 and DS1780) combine temperature sensing with other valuable features including

Table 1.2 Digital temperature sensors from Analog Devices

Type	Description
AD7414	SMBus/I²C digital temperature sensor in 6-pin SOT with SMBus alert and over temperature pin
AD7415	SMBus/I²C digital temperature sensor in 5-pin SOT
AD7416	Temperature-to-digital converter, I²C, 10-bit resolution, $-55\,°C$ to $+125\,°C$, $\pm2\,°C$ accuracy
AD7417	4-channel, 10-bit ADC with on-chip temperature-to-digital converter, I²C, $\pm1\,°C$ accuracy
AD7418	Single-channel, 10-bit ADC with on-chip temperature-to-digital converter, I²C, $\pm1\,°C$ accuracy
AD7814	10-bit digital temperature sensor in 6-lead SOT-23
AD7816	10-bit ADC, temperature monitoring only in an SOIC/μSOIC package

EEPROM arrays, real-time clocks and CPU monitoring. One more interesting feature of Dallas Semiconductor's temperature sensors is that they are expandable from 9 to 13 bits or user configurable to 9, 10, 11 or 12 bits resolution.

Dallas Semiconductor's DS1616 Temperature Data Recorder with the 3-Input Analog to Digital Converter adds the potential for three powerful external sensors to the base design of the DS1615 Temperature Data Recorder. It permits logging of not only the temperature, but also the humidity, the pressure, the system voltage, external temperature sensors, or any other sensor with the analog voltage output. The DS1616 provides all of the elements of a multi-channel data acquisition system on one chip. It measures the selected channels at user-programmable intervals, then stores the data and a time/date stamp in the nonvolatile memory for later downloading through one of the serial interfaces.

National Semiconductor also proposes some digital temperature sensors [32] with different temperature ranges from $-55\,°C$ up to $+150\,°C$: SPI/MICROWIRE plus the sign digital temperature sensor LM70 (10-bit) and LM74 (12-bit); the digital temperature sensor and the thermal watchdog with the two-wire (I^2C^{TM} Serial Bus) interface LM75 ($\pm3\,°C$); digital temperature sensors and the thermal window comparator with the two-wire interface LM76 ($\pm1\,°C$), LM77 ($\pm1.5\,°C$) and LM92 ($\pm0.33\,°C$). The sensors LM70, LM74 and LM75 include the delta-sigma ADC.

The window-comparator architecture of the sensors eases the design of the temperature control systems conforming to the ACPI (advanced configuration and power interface) specification for personal computers.

Another example of a digital output sensor is $+GF+$ SIGNET 2350 [33] with a temperature range from -10 to $+100\,°C$ and accuracy $\pm0.5\,°C$. The temperature sensor's digital output signal allows for wiring distances between sensor and temperature transmitter of up to 61 m. An integral adapter allows for the integration of the sensor and the transmitter into a compact assembly.

1.2 Pressure IC and Smart Sensors and Accelerometers

Similar to temperature sensors, pressure sensors are also very widely spread. In Europe, the first truly integrated pressure sensor was designed in 1968 by Gieles at Philips Research Laboratories [34], and the first monolithic integrated pressure sensor with digital (i.e., frequency) output was designed and tested in 1971 at Case Western Reserve University [35] as part of a programme addressing biomedical applications. Miniature silicon diaphragms, with the resistance bridge at the centre of the diaphragm and sealed to the base wafer with gold–tin alloy, were developed for implant and indwelling applications.

Pressure sensors convert the external pressure into an electrical output signal. To accomplish this, semiconductor micromachined pressure sensors use the monolithic silicon-diffused piezoresistors. The resistive element, which constitutes the sensing element, resides in the thin silicon diaphragm. Applying pressure to the silicon diaphragm causes its deflection and changes the crystal lattice strain. This affects the free carrier mobility, resulting in a change of the transducer's resistance, or piezoresistivity. The diaphragm thickness as well as the geometrical shape of resistors, is determined by the tolerance range of the pressure. Advantages of these transducers are:

- high sensitivity

- good linearity

- minor hysteresis phenomenon

- small response time.

The output parameters of the diffused piezoresistors are temperature dependent and require the device to be compensated if it is to be used over a wide temperature range. However, with occurrence smart sensors and MEMS, the temperature error can be compensated using built-in temperature sensors.

Most of today's MEMS pressure transducers produced for the automotive market consist of the four-resistor Wheatstone bridge, fabricated on a single monolithic die using bulk etch micromachining technology. The piezoresistive elements integrated into the sensor die are located along the periphery of the pressure-sensing diaphragm, at points appropriate for strain measurement [36].

Now designers can choose between two architectures for sensor compensation: the conventional analog sensor signal processing or digital sensor signal processing. The latter is characterized by full digital compensation and an error-correction scheme. With a very fine geometry, mixed-signal CMOS IC technologies have enabled the incorporation of the sophisticated digital signal processor (DSP) into the sensor compensator IC. The DSP was designed specifically to calculate the sensor compensation, enabling the sensor output to realize all the precision inherent in the transducer.

As considered in [37] 'as the CMOS process and the microcontroller/DSP technology have become more advanced and highly integrated, this approach may become increasingly popular. The debate continues as to whether the chip area and the circuit overhead of standard microprocessor designs used for this purpose will be competitive with less flexible (but smaller and less costly) dedicated DSP designs that can be customized to perform the specific sensor calibration function'.

The integrated pressure sensor shown in Figure 1.10 uses a custom digital signal processor and nonvolatile memory to calibrate and temperature-compensate a family of pressure sensor elements for a wide range of automotive applications.

This programmable signal conditioning engine operates in the digital domain using a calibration algorithm that accounts for higher order effects beyond the realm of most analog signal conditioning approaches. The monolithic sensor provides enhanced features that typically were implemented off the chip (or not at all) with traditional analog signal conditioning solutions that use either laser or electronic trimming. A specially developed digital communication interface permits the calibration of the individual sensor module via connector pins after the module has been fully assembled and encapsulated. The post-trim processing is eliminated, and the calibration and the module customization can be performed as an integral part of the end-of-line testing by completion of the manufacturing flow. The IC contains a pressure sensor element that is coprocessed in a submicron, mixed-signal CMOS wafer fabrication step and can be scaled to a variety of automotive pressure-sensing applications [37].

Now, let us give some state-of-the-art and industrial examples of modern pressure sensors and transducers. Major attention has been given to the creation of pressure sensors with frequency output in the USSR [38,39]. The first of them was based on the usage of VFC and had an accuracy up to 1%, the effective range of measuring

Figure 1.10   Monolithic pressure sensor (Reproduced by permission of Motorola)

frequencies 0–2 kHz in the pressure range 0–40 MPa. The second was founded on the usage of the piezoresonator. The connection of this device into a self-oscillator circuit receives a frequency signal, proportional to the force. The relation between the measured pressure p and the output frequency signal f is expressed by the following equation:

$$p = (f - f_0)/K_p; \; K_p = K_F \cdot S_{\text{eff}}, \tag{1.5}$$

where f_0 is the frequency at $p = 0$; f is the measurand frequency; K_p is the conversion factor of pressure-to-frequency; K_F is the force sensitivity factor; S_{eff} is the membrane's effective area.

The silicon pressure sensor based on bulk micromachining technology and the VFC based on CMOS technology was described in [40]. It has 0–40 kPa measuring pressure range, 280–380 kHz frequency output range and main error ±0.7%.

The Kulite company produces the frequency output pressure transducer ETF-1-1000. The sensor provides an output, which can be interfaced directly to a digital output. The transducer uses a solid-state piezoresistive sensing element, with excellent reliability, repeatability and accuracy. The pressure range is 1.7–350 bar, output frequency is 5–20 kHz, the total error band is ±2%. Other examples are pressure transducers VT 1201/1202 from Chezara (Ukraine) with 15–22 kHz frequency output range and standard error ±0.25% and ±0.15% accordingly.

The shining example of a sensor with $x(t) \to V(t) \to F(t)$ conversion is the pressure sensor from ADZ Sensortechnik GmbH (Germany). The IC LM 331 was used as the VFC. The output frequency of the converter can be calculated according to the following equation:

$$f_{\text{out}} = \frac{(U_{\text{in}} - U_{\text{offset}}) \cdot R6}{2.09V \cdot R4 \cdot R8 \cdot C6}, \tag{1.6}$$

where

$$U_{\text{in}} = P_{\text{abs}} \cdot 0.533 \frac{V}{\text{bar}} + 0.5V \tag{1.7}$$

The measuring range is 0–8.8 bar, the frequency range is 1–23 kHz.

The model SP550 from Patriot [41] is a rugged pressure transducer which provides full-scale output of 1–11 kHz. The output frequency can be offset to provide the output of any span between 1 kHz to 150 kHz. In this transducer the strain–guage signal is converted into the frequency via the precision monolithic VFC and the operational amplifier.

Geophysical Research Corporation has announced the Amerada® Quartz Pressure Transducer [42]. The transducer employs a rugged crystalline quartz sensor that responds to the stress created by the pressure. This response is in the form of a change in the resonant frequency created by the applied pressure. The pressure dependence of the sensor is slightly non-linear but is easily corrected during the calibration using a third-order polynomial function. In addition, the crystalline quartz is considered to be perfectly elastic which contributes to the excellent repeatability that is characteristic of this technology. Two additional quartz sensors are employed, one to measure temperature, the second acting as a stable reference signal. The temperature measurement is used for the dynamic temperature compensation of the pressure crystal while the reference signal is used as a stable timing base for the frequency counting. The pressure range is up to 10 000 psia, the accuracy up to ±0.02% FS [42]. The output pressure and temperature frequencies range between 10 kHz and 60 kHz.

The high-accuracy (0.01%) fibre-optic pressure transducers have been developed by ALTHEN GmbH by applying optical technology to resonator-based sensors [43].

Further development of microelectronic technologies and smart sensors has declined in the rise of high-precision (up to 0.01%) digital output pressure sensors and transducers. Some of which are described below.

The Paroscientific, Inc. Digiquartz® Intelligent Transmitter [44] consists of a unique vibrating quartz crystal pressure transducer and a digital interface board in the integral package. Commands and data requests are sent via the RS-232 channel and the transmitter returns data via the same two-way bus. Digital outputs are provided directly in engineering units with typical accuracy of 0.01% over a wide temperature range. The use of a frequency output quartz temperature sensor for the temperature compensation yields the achievable full-scale accuracy of 0.01% over the entire operating temperature range. The output pressure is fully thermally compensated using the internally mounted quartz crystal specifically designed to provide a temperature signal. All transmitters are programmed with calibration coefficients for full plug-in interchangeability. The Intelligent Transmitter can be operated either as a stand-alone standard output pressure sensor with the display, or as a fully integrated addressable computer-controlled system component. Transducers use crystalline quartz as the key sensing elements for both the pressure and the temperature because of its inherent stability and precision characteristics. The pressure-sensing element is a quartz beam, which changes frequency under the axial load. The transferred force acts on the quartz beam to give a controlled, repeatable and stable change in the resonator's natural frequency, which is measured as the transducer output. The load-dependent frequency characteristic of the quartz crystal beam can be characterized by a simple mathematical model to yield highly precise measurements of the pressure and pressure-related parameters. The output is a square wave frequency [44].

Other examples are intelligent pressure standards (series 960 and 970) [45]. In the 960 series, the pressure is measured via the change in the resonant frequency of the oscillating quartz beam by the pressure-induced stress. Quartzonix™ pressure standards produce the output frequency between 30 and 45 kHz and can achieve accuracy of $\pm 0.01\%$ FS. The precise thermal compensation is provided via the integrated quartz temperature sensor used to measure the operating temperature of the transducer. The 970 series uses a multi-drop, 9600 baud ASCII character RS-485 type interface, allowing a network of up to 31 transducers on the same bus. The output pressure measurement is user programmable for both the pressure units and update rate.

The resonant pressure transducer RPT 200 (frequency, RS-232/485 outputs) and the digital output pressure sensor RPT 301 (selectable output RS-232 or RS-485) series with $\pm 0.01\%$ FS accuracy are produced by Druck Pressure Measurement [46].

Another very popular silicon sensor from the mechanical signal domain is the accelerometer. The measurement of acceleration or one of its derivative properties such as vibration, shock, or tilt has become very commonplace in a wide range of products. The types of sensor used to measure the acceleration, shock, or tilt include the piezo film, the electromechanical servo, the piezoelectric, the liquid tilt, the bulk micromachined piezoresistive, the capacitive, and the surface micromachined capacitive. Each has distinct characteristics in the output signal, the development cost, and the type of the operating environment in which it best functions [47]. The piezoelectric has been used for many years and the surface micromachined capacitive is relatively new. To provide useful data, the first type of accelerometers require the proper signal-conditioning circuitry. Over the last few years, the working range of these devices has been broadened to include frequencies from 0.1 Hz to above 30 kHz.

Capacitive spring mass accelerometers with integrated electronics that do not require external amplifiers are proposed by Rieker Inc. These accelerometers of the Sieka series are available with analog DC output, digital pulse-width modulated, or frequency-modulated outputs [48].

The surface micromachined products provide the sensor and the signal-conditioning circuitry on the chip, and require only a few external components. Some manufacturers have taken this approach one step further by converting the analog output of the analog signal conditioning into a digital format such as a duty-cycle. This method not only lifts the burden of designing the fairly complex analog circuitry for the sensor, but also reduces the cost and the board area [47].

A very simple circuit can be used to measure the acceleration on the basis of ADXL202/210 accelerometers from Analog Devices. Both have direct interface to popular microprocessors and the duty-cycle output with 1 ms acquisition time [49]. For interfacing of the accelerometer's analog output (for example, ADXL05) with microcontrollers, Analog Devices proposes acceleration-to-frequency circuits based on AD654 VFC to provide a circuit with a variable frequency output. A microcontroller can then be programmed to measure the frequency and compute the applied acceleration [50].

1.3   Rotation Speed Sensors

There are many known rotation speed sensing principles and many commercially available sensors. The overwhelming majority of such sensors is from the magnetic signal domain (Hall-effect and magnetoresistor-based sensors) and the electrical signal domain (inductive sensors). According to the nature, passive and active electromagnetic sensors are from the frequency-time domain. Pulses are generated on its output. The frequency is proportional to the measured parameter. In such sensors the flow mesh is connected with the angle of rotation by the following equation:

$$\Psi = \Psi_m \cdot \cos\theta, \qquad (1.8)$$

then the induced emf in the sensor sensitive element is

$$e = -\frac{d\varphi}{dt} = \psi_m \cdot \sin\theta \frac{d\theta}{dt} = E_m \sin\theta \qquad (1.9)$$

and can be determined by the instantaneous frequency in any moment of time

$$f = \frac{1}{2\pi} \cdot \frac{d\theta}{dt} \qquad (1.10)$$

which is proportional to the instantaneous angular speed

$$\omega = \frac{d\theta}{dt} \qquad (1.11)$$

The current averaging of the angular speed ω on the interval h, measured in the units 2π represents the frequency of rotation and is determined by Steklov's function [51]:

$$n(t) = \frac{1}{2\pi h} \int\limits_{t-0.5h}^{t+0.5h} \omega(\tau)d\tau \qquad (1.12)$$

This equation reflects only a common idea of the current average of some function on its argument. Using the frequency measurement of the rotational speed the choice of mathematical expression should agree with experiment. In this case, the rotational speed and the angular velocity should be interlinked with the expression relevant to the physics of the observable process and the requirements of their measurement in a concrete system. For the description of rotational speed, it is expedient to use the expression of the flowing average:

$$n(t) = \frac{1}{2\pi h} \int\limits_{t-h}^{t} \omega(\tau)d\tau = \frac{1}{2\pi h} \int\limits_{0}^{h} \omega(t-\tau)d\tau \qquad (1.13)$$

The measuring frequency of the rotational speed is given by

$$n_x = f_x \cdot \frac{60}{Z} \qquad (1.14)$$

where Z is the number of modulation rotor's (encoder's) gradations.

For modern applications the rotation speed sensor should provide the digital or the quasi-digital output compatible with standard technologies. This means that the sensor and the signal-processing circuitry (the microcontroller core) can be realized in the same chip. An excellent solution in many aspects is when this signal is a square-wave output of an oscillator, the frequency of which is linearly dependent on the rotating speed and carries the information about it.

The semiconductor active position sensor of relaxation type designed by the authors together with the Autoelectronic company (Kaluga, Russia) can serve as an example [52,53]. It was developed on the basis of the crankshaft position sensor. Its principle of action is based on the effect of the continuous suppression of oscillations of the high frequency generator by passing each metal plate of the modulating rotor in front of the active sensor element and its subsequent resumption. Due to that, rectangular pulses with constant amplitude ($+V_{cc}$) are continuously formed on the sensor output. The frequency of these pulses is proportional to the rotating speed. If the metal tooth of the rotor-modulator comes nearer to the active element (generator coil), the logic level '1' ($+V_{cc}$) is formed as the sensor output. When the active sensor part appears between the teeth, the logic level '0' is formed as the output. Thus, the active sensor forms the pulsing sequence, the frequency of which is proportional or equal to the rotating speed. This sensor does not require any additional buffer devices for the tie-in measuring system and has a very easy interface to the microsystem. Moreover, the sensor meets the requirements of the technological compatibility with other components of the microsystem.

The circuit diagram of the Active Sensor of Rotation Speed (ASRS) is shown in Figure 1.11. It consists of a high frequency generator ($f = 1$ MHz), a sensing element (the generator coil), an amplifier, a voltage stabilizer and an output forming transistor with an opened collector [54]. The "chip & wire" technology was used in the sensor design, which combines the advantages of both monolithic and hybrid integrated technologies. All electronics was realized in a single chip, only the inductance, two resistors and the stabilitron were implemented in accordance with hybrid technology.

The ASRS is shown in Figure 1.12 and the sensor's output waveforms in Figure 1.13. The amplitude of the signal is constant and does not depend on temperatures and the direction of the rotation. The online time ratio $Q = 2$ (50% duty-cycle independent of the distance). But in a frequency range of more than 50 000 rpm, the pulse width will be increased.

The comparative features of modern non-contact sensors of different principles of function are shown in Table 1.3. Here sensors A5S07/08/09 are made by BR Braun (Germany), DZXXXX by Electro Corporation (USA); VT1855, OO020 by NIIFI (Penza, Russia); 4XXXX by Trumeter (UK); LMPC by Red Lion Controls (USA).

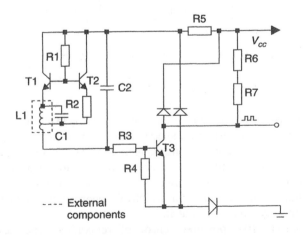

Figure 1.11   Circuit diagram of the Active Sensor of Rotation Speed (ASRS)

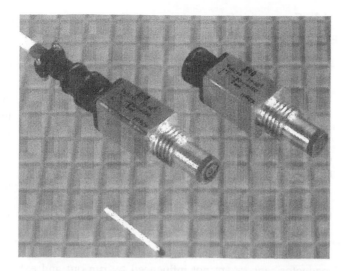

Figure 1.12 Active Sensor of Rotation Speed (ASRS)

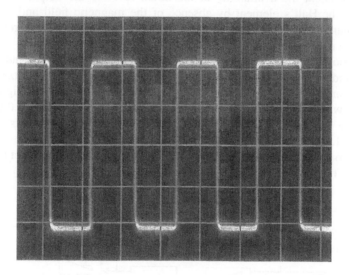

Figure 1.13 Waveforms of sensor's output horizontal scale 2.0 ms/div; vertical scale 5V/div

Active, magnetic and Hall-effect sensors are more suitable for the determination of the object status 'Stop' (a shaft is stopping). The advantage of active semiconductor sensors is the possibility of operation with the non-magnetic modulating rotor's teeth (steel, copper, brass, aluminium, nickel, iron). Therefore, the modulating rotor can be made of plastic and its teeth — of the metallized coating. It essentially raises the manufacturability and decreases the cost value. With the exception of the non-contact rotation speed sensing, such sensors can be used like an angular position sensor, a position sensor, a metallic targets counter and an end-switch. In addition, a smart sensor on this basis allows the measurement of the rotation acceleration.

Table 1.3 Comparative features of non-contact sensors of rotation speed

Sensors	Freq. range, kHz	Supply voltage, V	Current consumption, mA	Type
ASRS	0–50	4.5–24	7–15	active
A5S07	0.5–25	8–28	15+load current	Hall-effect
A5S08/09	0.5–25	8–25	15	Hall-effect
DZ375	0–5	4.5–16	20–50	magnetic
DZH450	0–5	4.5–30	20	Hall-effect
DZP450	1–10	4.5–16	50	Hall-effect
VT1855	0.24–160	27	3	inductive
OO 020	0.24–720	27	100	photo
4TUC	0.3–2	10–30	200	mag./inductive
4TUN	0.3–2	6.2–12	3	mag./inductive
45515	0.002–30	25	20	Hall-effect
LMPC	up to 10	9–17	25	mag./inductive

Active semiconductor sensors are not influenced by run-out and external magnetic fields in comparison with Hall-effect sensors. With Hall-effect sensors, it is necessary to take into account the availability of the initial level of the output signal between electrodes of the Hall's element by absence of the magnetic field and its drift. It is especially characteristic for a broad temperature range. A Hall-effect rotation speed sensor needs encoders with magnetic pole teeth.

Another good example of a smart sensor for rotation speed is the inductive position, speed and direction active microsensor MS1200 from CSEM (Switzerland) [55]. The output can switch up to 1 mA and is compatible with CMOS digital circuits, in particular with microprocessors. The frequency range is 0–40 kHz, the air-gap is 0–3 mm. The core is a sensor chip with one generator coil and two sets of detection coils (Figure 1.14).

The detection coils are connected in a differential arrangement, to reject the common mode signal. The sensor also includes an electronic interface, which is composed of

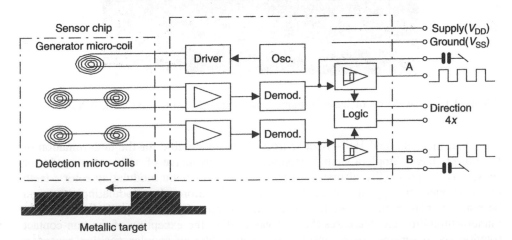

Figure 1.14 MS1200 functional block diagram (Reproduced by permission of POSIC S.A., Neuchatel, Switzerland)

a high frequency excitation for the generator coil and two read-out channels for the two sets of detection coils (channels A and B). The read-out electronics extract the amplitude variation of the high frequency signal due to the presence of a metallic target. The output stage is a first-order low-pass filter and a comparator. For a nominal target period of 2 mm, the outputs are two channels in quadrature (A quad B) as well as a direction signal and a speed signal (4X interpolation). It is composed of two silicon chips, one for the integrated microcoil and the other for the integrated interface circuit. The sensor produces a two-channel digital output, as well as a direction signal.

1.4 Intelligent Opto Sensors

Next, we shall examine a technique of delivering the output from optical (light) sensors into the frequency–time (quasi-digital) domain. Light is a real-world signal that is often measured either directly or used as an indicator of some other quantity. Most light-sensing elements convert light into an analog signal in the form of the current or the voltage, then a photodiode current can be converted into the frequency output. Light intensity can vary over many orders of the magnitude, thus complicating the problem of maintaining resolution and signal-to-noise ratio over a wide input range. Converting the light intensity to a frequency overcomes limitations imposed on the dynamic range by the supply voltage, the noise and the ADC resolution.

One such device is a low-cost programmable silicon opto sensor TSL230/235/245 from Texas Instruments with a monolithic light-to-frequency converter [56]. The output of these devices is a square wave with a frequency (0–1 MHz) that is linearly proportional to the light intensity of the visible and short infra red radiation. Additionally the devices provide programming capability for the adjustment of the input sensitivity and the output scaling. These capabilities are effected by a simple electronic technique, switching in different numbers of the 100 elements of the photodiode matrix. For costs reasons, the low-cost microcontroller with a limited frequency range may be used for the frequency-to-digital converter due to the output scaling capability. Options are an undivided pulse train with the fixed pulse width or the square wave (50% duty-cycle) divided by 2, 10 or 100 outputs. Light levels of 0.001 to 100 000 μW/am^2 can be accommodated directly without filters [56].

Since the conversion is performed on-chip, effects of external interference such as noise and leakage currents are minimized and the resulting noise immune frequency output is easily transmitted even from remove locations to other parts of the system. The isolation is easily accomplished with optical couples or transformers.

Another interesting example is the integrated smart optical sensors developed in Delft University of Technology [57,58]. Integrated on-chip colour sensors have been designed and fabricated to provide a digital output in the IS2 bus format. The readout of photodiodes in the silicon takes place in such a way that pulse series are generated with the pulse frequency proportional to the optical intensity (luminance) and the duty-cycle to the colour (chrominance). The colour information is obtained using the wavelength dependence of the absorption coefficient in the silicon in the optical part of the spectrum, so no filters are required. The counters and the bus interface have been realized in a bipolar and CMOS version with enhanced resolution which is being investigated.

1.5   Humidity Frequency Output Sensors

Frequency–time-domain humidity and moisture sensors can be created based on humidity-capacitance-frequency (or duty-cycle) converters. Relative humidity sensors HS1100/HS1101 from HUMIREL (USA) are based on a unique capacitive cell. Together with two types of frequency (duty-cycle close to 50%) output circuits, these are sensors with $x(t) \rightarrow C(t) \rightarrow F(t)$ conversion (third group in Figure 1.4). The circuits are based on the IC TLC555. Though these timers are not precise, the conversion error does not exceed 1%. The typical frequency range is 5978–7285 Hz. Based on the rugged HS1101 humidity sensor, HF3223/HTF3223 is a dedicated humidity transducer designed for OEM applications where reliable and accurate measurement is needed. The direct interface with a microcontroller is made possible by the module's linear frequency output (8030–9560 Hz) for 10–95 RH (%) measurements [59].

E + E ELEKTRONIK (Germany) manufactures a humidity/temperature transmitter with frequency output (EE05, EE 25 series) [60]. It provides a pulsed signal for both the humidity and the temperature. Every microprocessor system is able to read these data by simply counting the pulses without expensive A/D converting. The measuring range is 0–100% RH, the frequency range is 62.3–47.1 kHz and 12.5–9.4 kHz dependent on the type and the accuracy is ±2% RH (for EE 25 series).

Another humidity sensor was fabricated by Galltec (Germany). It is also the sensor with $x(t) \rightarrow C(t) \rightarrow F(t)$ conversion. One application circuit with the frequency output for FE09/1, FE09/2 and FE09/4 humidity sensing elements is designed on the discrete components (5–95% RH corresponding to 54–47 kHz). The second application circuit with the frequency output for the same sensing elements is based on the IC 555 (5–95% RH corresponding to 33–27 kHz and 3–2 kHz for FE09/1000 sensing element).

1.6   Chemical and Gas Smart Sensors

An academic/industrial UK LINK project is currently developing the handheld electronic nose (H^2EN) — an array of sensors simulating the human olfactory response. The approach described in [61] places the resistive sensor in the feedback loop of a 'digital' RC oscillator, the output of which is a square wave with a frequency inversely proportional to the sensor resistance. The frequency measurement technique used (period counting) counts the number of cycles of the internal FPGA oscillator module n_o over a period equal to a fixed number of cycles of the sensor oscillator n_s. The current prototype H^2EN uses an embedded ELAN SC400 486SX microcontroller. In order to reduce the interference due to noise the acquired data are averaged by increasing n_s until the noise performance becomes acceptable. This approach achieves two objectives: firstly, random errors due to noise tend to be minimized and secondly the resolution of the final measurement is improved. Both are achieved at the expense of the measurement speed.

The sensors constituting the array are selected for their chemical affinities and are typically based on chemisorbing polymer films. Many of the following sensors can be used, and a serial polling of each sensor reading creates outputs. In an ideal array response each output corresponds to only one analysis or chemical compound. One

Table 1.4 Industrial frequency–time-domain and digital sensors on sensors web portal (http://www.sensorsportal.com)

Sensor/transducer	Range	Frequency output	Accuracy	Relative error,%
		Pressure		
ETF-1-1000 (Kulite GmbH)	1.7...350 bar	5...20 kHz	*	±2%
ADZ Sensortechnik GmbH	0...8.8 bar	1...23 kHz	*	*
VT 1201/1202	0.5...180	15...22 kHz		±0.25%
CHEZARA (Ukraine)	(60) MPa		*	(0.15%)
Druck (Germany), RT200, RT301	*	Digital	*	0.01%
Amerada, QPT16K177C, QPT16K150C	16 000 psia	10...60 kHz		±0.02%
QPT10K150C	10 000 psia	10...60 kHz	*	± 0.02%
Patriot, SP550	10 000 (20 000) psia	1...11 kHz	*	*
Keyence, AP30	0...1.000 MPa	Digital		0.2%
AP40	0...1 MPa	Digital	*	0.2%
ALTHEN, 8000 (8DP, 8WD, 8B)	*	Digital/frequency	*	0.01%
Paroscientific Inc., Digiquartz Model 710	15...40 000 psia	Digital	*	0.01%
Pressure Systems, Quartzonix, Series 960, Series 970	15...500 psia	30...45 kHz		0.01%
	0...115 kPa (0...3333 kPa)	Digital	*	0.01%
		Temperature		
SMT 160-30 (SMARTEC, The Netherlands)	−45...+130°C	1...4 kHz	±0.7°C	±0.54%
DS1821 (DALLAS, USA)	−55...+125°C	Digital		±1%
AD7816/7817/7818 (Analog Device)	−55...+125°C	Digital	±2.0°C	*
National Semiconductor, LM77	−55...+125°C	Digital (9 bits)	±3.0°C	*
THSENS-F (Hungary)	−50...+150°C	400...1600 kHz	±2.0°C	*
SBE, Model SBE8, Model SBE 3F	0...+30°C	0.1...200 Hz	±0.01°C	
	−5...+35°C	2...6 kHz	±0.001°C	*
Slope Indicator (USA), VW sensors	−45...100°C		±0.1°C.	*
		Humidity		
HS1100/HS1101 (HUMIREL, USA)	0...100% RH	7285...5978 Hz	*	> ±1%
EE 05/25 (E + E ELEKTRONIK GmbH)	0...100% RH	62.3...47.1 kHz	±2% RH	
		12.5...9.4 kHz		*
FE09/1, FE09/2, FE09/4, FE09/1000		30...300 kHz		
(Galltec GmbH)	0...100% RH	3...30 kHz	±1.5% RH	*

(continued overleaf)

Table 1.4   (*continued*)

Sensor/transducer	Range	Frequency output	Accuracy	Relative error,%
		Light		
TSL230, 235, 245	300...1100 nm		*	
(Texas Instruments, USA)	+infrared range	0...1.1 MHz		±5%
		Rotation Speed		
A5S07/08/09 BR Braun GmbH	*	0.5...25 kHz	*	*
BES516-371/324/325/356		0...3000 Hz		
BES516-349/384	*	0...5000 Hz	*	*
BALLUFF (Germany)				
MS1200 CSEM (Switzerland)	*	0...40 kHz	*	*
14.3862 (Russia)	*	20...830 Hz	*	*
		Flow		
VR00, VR10, VR25 SIKA	15...1000 l/min	10...1000 Hz	*	±3%
(Germany)				
Hayward, Model 2000	0.15...10 m/sec	42 Hz/m/sec	*	±1%

* — Information is not available.

such solid-state SAW sensor is described in [62]. This sensor is an uncoated, high-Q piezoelectric quartz crystal with a natural resonating frequency of 500 MHz. Its surface temperature is controlled by a small thermoelectric element that cools the surface to promote the vapour condensation and then heats it for cleaning between analyses.

The added mass of the analyte condensing on the crystal's surface lowers the vibration frequency in direct proportion to the amount of the condensate. This frequency is mixed with a reference frequency, and the intermediate frequency (typically 100 kHz) is counted by a microprocessor.

The acoustic gas sensor is described in [63]. In this sensor, the sound velocity is continuously measured with high resolution in a gas-filled cell, by controlling the frequency of an oscillator via the transit time of the sound between an ultrasound transmitter and a receiver element. The circuitry for control, signal processing and communication is based on the microcontroller PIC-17C44. This microcontroller can handle frequency input signals directly, and also provides analog output signals by means of pulse width modulation (PWM).

Chemical signal domain sensors can be used not only for measuring different chemical quantities and compositions of mixed gases, but also for measuring 'non-chemical' quantities such as the rotation speed [64], for which electrical and magnetic signal domain sensors are usually used. Electrochemical oscillations have often been observed in iron immersed in a solution containing phosphoric acid and hydrogen peroxide. This oscillation has been interpreted by the cross-linkage between electrochemical reactions and mass transport processes in the vicinity of an electrode. Therefore, the oscillation frequency is expected to reflect the flow rate around the electrode. A rod of carbon steel S50C was mounted in the rotating disc apparatus. The intersection of the rod rotated concentrically. In a mixture of 0.5 M phosphoric acid and 1.5 M hydrogen peroxide the highly stable electrochemical oscillation appeared. The amplitude of the potential change was about 600 mVp-p. The oscillating frequency was a sensitive function of the rotation rate in the range of 100 to 2000 rpm.

Summary

The huge number of frequency–time-domain sensors is certainly not covered by this review. However, from this survey it is possible to draw the following conclusions. The rapid development of microsystems and microelectronics in a whole promotes the further development of different digital and quasi-digital smart sensors and transducers. Today there are frequency–time-domain sensors practically for any physical and chemical, electrical and non-electrical quantities. The obvious tendency of accuracy increasing up to 0.01% and above is observed. These devices work in broad frequency ranges: from several hundredth parts of Hz up to several MHz. The extension of their 'intelligent' capabilities including intelligent signal processing is traced. The wide distribution of these sensors has accorded converters of different parameters (voltage, current, capacity etc.) into frequency.

The process of miniaturization boosts the creation of multichannel, multifunction (multiparameter) one-chip smart sensor and sensor arrays. The basic characteristics of the majority of described sensors are listed in Table 1.4. More detailed information can be obtained from Sensors Web Portal (http://www.sensorsportal.com).

2

CONVERTERS FOR DIFFERENT VARIABLES TO FREQUENCY-TIME PARAMETERS OF THE ELECTRIC SIGNAL

Apart from the various types of frequency sensors considered below there is another big group of analog sensors with current, voltage or resistance output. These are, for example, thermocouples, potentiometric sensors etc. When using the frequency signal as the sensor's informative parameter, it is expedient to use voltage (current)-to-frequency converters of the harmonious or pulse signal in the measuring channel. Such converters are frequently used in practice due to their good performance: linearities and the stability of transformation characteristics, accuracy, frequency ranges, manufacturability and simplicity.

2.1 Voltage-to-Frequency Converters (VFCs)

The operational principle of majority VFC consists in alternate integration of input voltage and generation of pulses when the integrator's output voltage equals to the reference voltage. VFC with continuous integration have special advantages. Such converters do not require additional time for returning the integrator to an initial condition at the beginning of a new conversion cycle.

Initially such converters were developed based on magnetic materials with a rectangular hysteresis loop. These were the elementary magneto-transistor multivibrators with input frequency directly proportional to the input voltage [65,66]. Various modifications with improved characteristics and possibilities [67–69] were further developed. For example, a pulse-width modulator with constant amplitude of output pulses and direct proportionality of the duty factor to input voltage has been described in [67]. A magneto-transistor multivibrator with high sensitivity (1000 Hz/mV), 1000 Ω input resistance for usage in VFCs for sensors such as thermocouples, resistive thermometers and strain gauges [68] has been developed. It has a voltage range 0–20 mV at nonlinearity of transformation characteristic ±0.4%. The high initial frequency of the input signal 250 kHz essentially reduces the transient time and the drive circuit's inductance. The distinctive feature of the multivibrator offered in [69] is its use for the linear transformation of positive as well as negative voltage into frequency. However, the use of VFCs with magnetic elements has not received a wide distribution for the precise transformation of the voltage (current) into a pulse-frequency signal because of the strong dependence of the core's parameters on temperature. The suggested circuits for the temperature compensation were too complex and did not allow the complete exclusion of temperature influences.

The development of integrating VFCs based on symmetric controlled multivibrators was made in parallel. They differed by simplicity and high technical and dynamic characteristics [70,71]. Working in multichannel code-pulse remote metering systems with frequency output sensors in the industrial environment in temperatures up to +50 °C, VFCs reliably carried out linear transformation of the voltage 4–45 V into frequency signals 1.8–20.0 kHz. They guaranteed nonlinearity of the transformation characteristic not worse than ±0.2% and relative temperature error up to ±0.5 % [72,73]. In the patent [71] it is described how the authors removed the nonlinearity error caused by the non-identity of the two driving current circuits, providing the linear charge (or discharge) of integrating capacitors in the multivibrator. It was achieved due to the usage in the multivibrator of one driving current transistor instead of two.

It has been demonstrated, that by using stricter requirements to the conversion accuracy, it is expedient to use similar integrating elements, for example, the charge of the capacitor and the compensation of it by the discharge impulse with constant amplitude. Therefore, the most acceptable for usage in integrated VFCs were:

— Integration with the help of the capacitor (change of the multivibrator's frequency by the change voltage, the current or the resistance). The range is 10–100% and the basic error is 0.5%.

— Charge of the capacitor by the direct current proportional to the measurand and the fast discharge by achievement of the threshold level (the range is 0–100%, the fiducial error is 0.3%).

— Charge and discharge of the capacitor by the direct current proportional to the measurand (the range is 0–100%; the fiducial error is 0.1%).

Creation of operational amplifiers and other elements of microelectronics have opened a way to design more perfect integrators and the VFC on the basis of new principles of integration.

VFCs of direct transformation with open structures, and VFCs of the compensatory type with feedback [74–76] can be defined. Among the open-type VFCs greatest distribution has been received integrating converters based on the cyclic charge during time T_0, and the discharge during $T_{discharging}$ of the integrator's integrating capacity (Figure 2.1a).

The main source of methodical error is the time off T_0 during which the integrator is prepared for a new cycle of transformation $V_x \rightarrow f_x$ (i.e. comes back to an initial state). Reduction of this error requires the inequality satisfiability $T_0 \ll T_{discharging}$, and its suppression $T_0 = 0$. A more radical method of error suppression is to use two alternatively working integrators (Figure 2.1b), or the usage of the integration and the

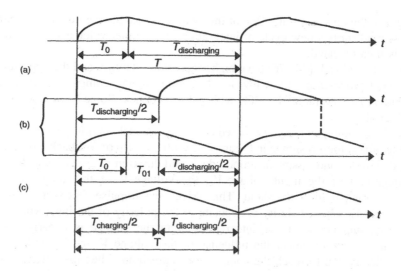

Figure 2.1 Charge and discharge diagrams of integrating capacitors

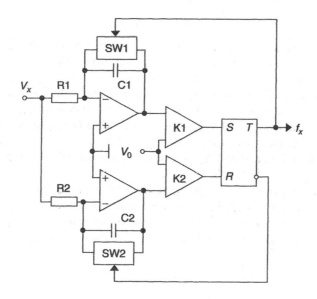

Figure 2.2 Two-integrator VFC

'disintegration' (Figure 2.1c). In these cases, the methodical error will be completely suppressed.

A simple VFC containing two integrators based on operational amplifiers is shown in Figure 2.2.

Each of the integrators is working alternatively, continuously–cyclically without dividing pauses between cycles, due to the RS-trigger with comparators on inputs. The trigger is nulled each time an input voltage of one of the integrators equals the stable reference voltage V_0. Owing to the comparator's output the signal will toggle the

RS-trigger in the opposite state. One of the switch discharges the integrator's capacitor and the second starts a new cycle of integration of the input voltage V_x–the described process is then repeated.

VFCs with feedback [74–76] have higher accuracy, stability and the opportunity for sensor signal processing with a low level of input signal, compared with open-type VFCs. The most preferable is a VFC with pulse feedback. Some basic types [76] will be considered below.

The VFC including an integrator based on an operational amplifier, a comparator and impulse feedback loop is shown in Figure 2.3. With the help of feedback loop the pulses with stable volt-second square S_0, amplitude V and duration τ_0 are periodically formed.

At integration of the input voltage V_x, the voltage V of the integrator's output reaches the comparator's threshold. This drives the impulse feedback loop, which forms a pulse V_0 whose polarity is opposite to the polarity of the input voltage. With the feedback impulse S_0 on the integrator's input the voltage is linearly increasing during time t_0. Further, under the influence of the voltage V_x the integrator's output voltage is decreasing to reach the comparator's threshold. Then the conversion cycle of $V_x \rightarrow f_x$ is repeated again. The VFC's output frequency while forming the impulse feedback is.

$$f_x = \frac{R_0}{R} \cdot \frac{1}{V_0 \cdot \tau_0} \cdot V_x \qquad (2.1)$$

For an ideal integrator, the output frequency is proportional to V_x and does not dependent on the feedback pulse form. Only the stability of the volt-second square S_0 is important, which is not a difficult technical task.

Another version of a VFC with impulse feedback, but with separate formation of amplitude V_0 and the duration τ_0 is shown in Figure 2.4. It contains an integrator and a comparator based on operational amplifiers A1 and A2; a D-trigger is applied, a clock then inputs reference pulses of frequency $f_{0min} = 2.5 f_{xmax}$, the voltage reference, the analog switch and resistors R and R_0 of integrating circuits of the input and reference voltage source.

This VFC functions almost like the previous VFC (Figure 2.3) except that at the comparator operation, the input signal operates on the D-trigger's input, and at receipt of the first impulse of the reference frequency f_0 on the trigger's clock input the last one is toggled and produces the driving pulse for the analog switch. Thus by the

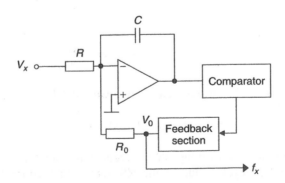

Figure 2.3   VFC with impulse feedback loop

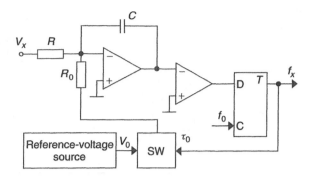

Figure 2.4   VFC with impulse feedback and separate forming of amplitude V_0 and duration τ_0

integrator's input through the resistor R_0 the reference voltage V_0 is connected. After the capacitor has been discharged, the comparator E_0 is returned to an initial stage by the pulse of frequency f_0. The D-trigger is also returned to the initial stage and disconnected from the integrator reference voltage V_0.

There are the following sources of the VFC conversion error:

1.   inaccuracy and instability of the R_0/R ratio

2.   inaccuracy of integration

3.   instability of the comparator's threshold during one conversion cycle

4.   absence of synchronization of pulses f_0 with the moments of the comparator operation

5.   instability of the reference voltage V_0

6.   instability of the time interval τ_0.

The use of the appropriate circuitry and technological measures may reduce the errors of $V_x \rightarrow f_x$ conversion listed above.

Further modification of the circuits considered above, which allows reduction of the nonlinearity error and the error due to synchronization absence between the moments of the comparator operation and the beginning of the feedback impulse forming is shown in Figure 2.5. It is achieved by the additional J–K trigger, the coincidence circuit, the frequency divider with division factor n and increasing the reference frequency f_0 by n times.

Due to the increase in frequency f_0 by n times and the use of the frequency divider with the division factor n on the J–K trigger's input the feedback impulse duration remains constant and equal to:

$$\tau_0 = \frac{1}{f_0}. \tag{2.2}$$

However, the maximal delay of the D-trigger operation is decreased and does not exceed $1/nf_0$. Due to this, the VFC accuracy is increased.

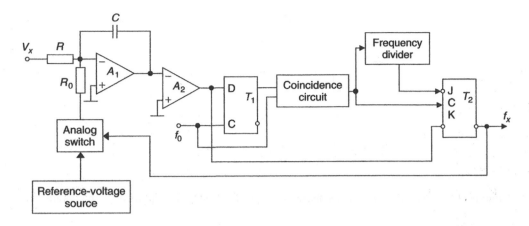

Figure 2.5   VFC modification with impulse feedback

Let us consider principles of operation, features and technical performances of some interesting integrating VFCs with a pulse feedback.

The full-range VFC is based on the balanced charge and discharge of the integrator's capacitor. It was created for the functionality extension of the digital frequency counter for positive voltage, current and capacitance measurements [77,78]. Its circuit is shown in Figure 2.6 and comprises an integrator, a comparator, a digital univibrator based on a pulse counter and a quartz generator, a signal conditioner, a source of stable current i and a high-speed transistor differential current switch.

The conversion is performed continuously–cyclically without dividing pauses. During the first part of the cycle the input current I_x is integrated and during the second, the difference of currents I_x and i. Its duration is equal to the pulse width T_x of the output frequency f_x. The distinctive feature of the method [78] is the equality of quantity of electricity on the integrator's input and that consumed by the source of current through the high-speed differential switch, which is carried out for some VFC cycles because

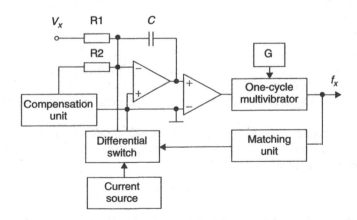

Figure 2.6   Full-range VFC based on balanced charge and discharge of integrator's capacitor

of uncertainty of the time delay of the digital univibrator operation within the limits of 0.125 µs. The measuring conversion's equation may be deduced as follows.

The charge Q_1 acting during the time T_x to the integrator's input from the source of voltage V_x is equal to:

$$Q_1 = I_x \cdot T_x = \frac{V_x \cdot T_x}{R} \tag{2.3}$$

and charge Q_2 consumed from the integrator's input by $q = i\tau$ equals:

$$Q_2 = N \cdot q = N \cdot i \cdot \tau, \tag{2.4}$$

where N is the number of positive pulses on the univibrator's output; τ is the pulse duration.

From the condition of equality $Q_1 = Q_2$, reflecting the principle of the VFC operation, it follows that

$$f_x = \frac{N}{T_x} = \frac{Q_2}{q \cdot T_x} = \frac{I_x}{q} = \frac{V_x}{q \cdot R} = \frac{V_x}{i \cdot \tau \cdot R} = k \cdot V_x, \tag{2.5}$$

where k is the conversion factor of the VFC. Hence the conversion $V_x \to f_x$ is linear, and $\delta f_x = \delta k$.

Taking into account the following values from the circuit $R = 10^7$ Ω, $i = 2 \times 10^{-4}$ A, $\tau = 5 \times 10^{-7}$ s the following equation for the direct readout is obtained:

$$f_x(\text{Hz}) = 1000 \cdot V_x(\text{V}); \; V_x(\text{V}) = f_x(\text{kHz}). \tag{2.6}$$

The formula for current conversion is obtained in a similar way:

$$f_x(\text{Hz}) = 10^{10} I_x(\text{A}); \; I_x(\text{nA}) = 0.1 \cdot f_x(\text{Hz}). \tag{2.7}$$

As follows from Equation (2.5) the main VFC error (instability) is determined by the elements of its pulse compensating feedback. The instability of the integrator's capacitance, threshold of the comparator operation and the time delay of the digital univibrator practically do not influence the VFC operation.

The VFC can work in a wide range of positive measuring voltages 1 mV–1000 V without any switching as well as currents in limits from 10^{-10} up to 10^{-4} A and higher with the help of additional shunts. Together with a digital frequency counter and time of measurement $T_x = 1$ s it is possible to measure the voltage of positive polarity with the frequency error not exceeding $10^{-3} f_x \pm 0.1$ Hz. This VFC can be used similarly to the controlled by voltage full range 10–10^6 Hz linear generator.

A voltage-to-frequency converter is shown in Figure 2.7. It realizes the voltage-to-frequency conversion according to the method of binary integration [79]. At relative simplicity and minimum hardware, this VFC achieves high accuracy for a wide range of input voltages V_x.

This VFC contains an integrator based on an operational amplifier, a control switch, a quartz generator, a threshold device and a time interval shaper of constant duration 0.1 s.

The $V_x \to f_x$ conversion is carried out in two steps. At the first step (duration T_1), only the voltage V_x of negative polarity is integrated, and at the second step (duration T_0) both the voltage V_x and the reference voltage V_0 are integrated simultaneously.

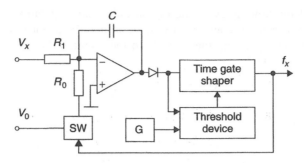

Figure 2.7   VFC based on method of binary integration

During step T_1 the input signal acts through the resistor R_1 on the inverting integrator's input, the control switch will be closed. Thereafter the output voltage of the integrator is increased, and the greater V_x, the greater its steepness and less the duration T_1. As soon as it reaches the threshold value the threshold device acts and drives the impulses of the quartz generator on the time interval shaper's input. By the pulse propagation through the shaper's divider the first voltage transition from '1' to '0' logical level on the shaper's output opens the control switch. The time interval T_0 which is forming begins with the same moment, during which voltage V_0 is acting on the inverting integrator's input through the resistor R_0. Its polarity is opposite but the amplitude is much higher. As a result, the output voltage of the integrator will be linearly decreasing. After the second transition from '0' to '1' the forming of the time interval T_0 is finished, the switch is closed, the pulse propagation of the quartz generator into the pulse shaper is stopped and the VFC is again in the mode of integration, only the input voltage V_x. Thus the averaging periodic process of conversion is realized in the VFC.

Idealizing the operation of the VFC's units, the steady state can be described by the following equation:

$$\int_0^{T_1} k_1 V_x \mathrm{d}t - \int_{T_1}^{T} (k_2 V_0 - k_1 V_x)\, \mathrm{d}t = 0, \tag{2.8}$$

where k_1 and k_2 are the integrator scaling for the measuring and compensating signals accordingly; T is the working cycle of the VFC during both conversion steps. Consequently $T = T_1 + T_0$.

If the voltage V_x *is constant* during the readout period, Equation (2.8) can be written:

$$k_1 V_x T - k_2 \int_{T_1}^{T_1+T_0} V_0\, \mathrm{d}t = 0 \tag{2.9}$$

Otherwise the integral in Equation (2.8) may be replaced with the product $V_x T$, where V_x is the average input voltage during the time $(T_1 + T_0) = T$. After simple transformation, we obtain the linear dependence of the measuring transformation:

$$f_x = \frac{1}{T} = \frac{k_1 \cdot V_x}{k_2 \cdot S_0} = S \cdot V_x, \tag{2.10}$$

where

$$S_0 = \int_0^{t_0} V_0 \, dt \qquad (2.11)$$

is the volt-second square of the feedback impulse,

$$S = \frac{k_1}{k_2 \cdot S_0} = \frac{R_0}{R_1 \cdot S_0}; \quad k_1 = \frac{1}{R_1 \cdot C} \cdot k_2 = \frac{1}{R_0 \cdot C} \qquad (2.12)$$

The analysis shows that the function of the measuring conversion of the VFC is determined by the size of the volt-second square of feedback impulses and the ratio of the summing resistance on the integrator's input. It is obvious that the VFC accuracy at first approximation depends on the stability of the specified parameters. Thus the advantage of closed compensation VFC circuits above open circuits.

The above-considered VFC has a high enough conversion factor (10 kHz/V) for the measuring interval 0.1 s and can be increased. The resolution is 1 mV, and the conversion error is 0.05%.

Another voltage-to-frequency converter [80] in comparison with that considered above converts the voltage V_x to the output frequency f_x as a result of continuous tracing compensation by two input V_x and compensating $V_{compensated}$ voltages which act in the feedback loop as a result of the return transformation of the output frequency f_x. This converter differs from other such VFCs in that the impulse shaper of the stable volt-second square is entered into the feedback loop. Due to this, the reliability and accuracy of conversion are essentially increased.

The circuit of such a VFC is shown in Figure 2.8 and consists of the subtracting unit carrying out the function of the comparison element of V_x and $V_{compensated}$ voltages and forming the voltage difference $\Delta V = V_x - V_{compensated}$ and two circuits–direct and indirect. The direct circuit consists of an amplifier of the direct current and a controlled generator. Usually in the direct circuit, high-sensitivity and less stable converters are used. The feedback loop, in which as a rule, highly stable converters of homogeneous size are used, consists of the impulse shaper of the stable volt–the second square, the voltage divider, the integrator, the pulse-amplitude modulator and the fixing element.

The $V_x \to f_x$ conversion is realised by the following way. The input voltage of undercompensation ΔV_x on the subtracting unit's output is transformed in the direct circuit into the output frequency f_x. In the steady state the frequency f_x of the controlled generator is proportional to the voltage $V_{compensated}$. The integrator realizes the integral from the sum of two voltages: V_1 from the pulse shaper's output and V_2 from

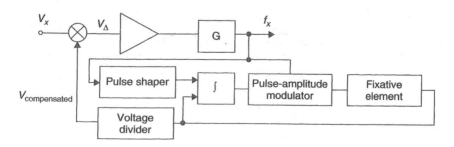

Figure 2.8 VFC of tracing compensation type

the fixing element's output. The output voltage of the integrator is acting once per period T_x through the pulse-amplitude modulator to the fixing element. Both of these elements form the negative feedback loop. As a result, the voltage V_2 on the fixing element's output is equal to the average voltage during the period T, acting from the pulse shaper's output of the stable volt–the second square and is equal to:

$$V_2 = \frac{1}{T} \cdot \int_0^T \cdot V_1 \, dt \tag{2.13}$$

Thus, on the pulse shaper's output there are pulses stabilized on the amplitude V_0 and the duration τ_0. As a result, the voltage $V_{\text{compensated}}$ on the voltage divider's output within the size of undercompensation ΔV can be accepted to be equal to the converted voltage V_x. In view of the transfer factor k of the voltage divider, we will receive the required linear dependence between the frequency f_x and the input voltage V_x:

$$f_x = \frac{1}{T_x} = V_x \cdot \frac{1}{k \cdot V_0 \cdot \tau_0} \tag{2.14}$$

This VFC of the tracing compensation type provides increased reliability and accuracy of measuring instruments and ADC, working in data acquisition systems.

Let us deduce the basic equations describing a static VFC at which the balance of the measuring circuit is incomplete. We shall notice that due to the presence of the undercompensation voltage $\Delta V = V_x - V_{\text{feedback}} = f_x / k_{dc}$ the balance of the circuit is provided, though ΔV is a source of the static error, the value of which can be reduced up to a minimum [81,104].

Let us consider the common block diagram of a static VFC (Figure 2.9). We shall accept that the subtracting unit is the converter of the circuit of the direct conversion of ΔV and we shall deduce some basic equations describing its operation and characteristics [81].

For the state stage taking into account the linearity of characteristics of measuring transformations and conversion factors of the subtracting device k_{sub}, factors of direct k_{dc} and indirect k_{id} conversions, the following equations are valid:

$$\Delta V = k_{\text{sub}}(V_x - V_{\text{feedback}}) \tag{2.15}$$

$$f_x = k_{dc} \cdot \Delta V = k_{dc} \cdot k_{\text{sub}}(V_x - V_{\text{feedback}}) \tag{2.16}$$

$$V_{\text{feedback}} = k_{ic} \cdot f_x \tag{2.17}$$

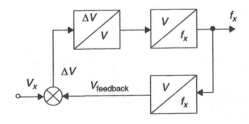

Figure 2.9   Common block diagram of a static VFC

Having carried out transformations and solved the equation relative to f_x we have:

$$f_x = \frac{k_{sub} \cdot k_{dc}}{1 + k_{sub} \cdot k_{dc} \cdot k_{ic}} V_x \qquad (2.18)$$

Hence, the conversion factor of the VFC or sensitivity of balancing is

$$k_{VFC} = \frac{k_{sub} \cdot k_{dc}}{1 + k_{sub} \cdot k_{dc} \cdot k_{ic}}, \qquad (2.19)$$

with value determined by units of measurements f_x, V_x and it is in $(1 + k_{sub}k_{dc}k_{ic})$ times less than k_{dc} but for this reason as will be shown below, the conversion accuracy, the voltage range and the signal-to-noise ratio will be increased at the same time.

For the case when $k_{sub}k_{dc}k_{ic} \gg 1$

$$k_{VFC} = \frac{1}{k_{ic}} \qquad (2.20)$$

It specifies that metrological and dynamic characteristics of the compensation VFC are completely determined by similar characteristics of the indirect conversion. After taking the logarithm, the differentiation and the transition to the final increments we have:

$$\delta k_{VFC} = -\delta k_{ic} \qquad (2.21)$$

For the case when the condition $k_{sub}k_{dc}k_{ic} \gg 1$ is not valid, we have:

$$\ln k_{VFC} = \ln k_{sub} + \ln k_{dc} - \ln (1 + k_{sub}k_{dc}k_{ic}), \qquad (2.22)$$

and after taking the logarithm, differentiation and transition to the final increments

$$\delta k_{VFC} = \frac{\delta k_{sub}}{1 + k_{sub}k_{dc}k_{ic}} + \frac{\delta k_{dc}}{1 + k_{sub}k_{dc}k_{ic}} - \delta k_{ic}\frac{k_{sub}k_{dc}k_{ic}}{1 + k_{sub}k_{dc}k_{ic}} \qquad (2.23)$$

For the most typical case, when $k_{sub} = 1$, we have:

$$\delta k_{VFC} = \frac{\delta k_{sub}}{1 + k_{dc}k_{ic}} + \frac{\delta k_{dc}}{1 + k_{dc}k_{ic}} - \delta k_{ic}\frac{k_{dc}k_{ic}}{1 + k_{dc}k_{ic}} \qquad (2.24)$$

Hence, the multiplicative component of the error in the circuit of the direct conversion has been decreased $(1 + k_{dc}k_{ic})$ times. Moreover, if errors δk_{dc} and δk_{ic} are caused by the same reasons they are subtracted.

Other characteristics of the VFC are the following:

- The balancing depth $(k_{dc}k_{ic})$ or the loopback amplification is equal:

$$k_{dc}k_{ic} = \frac{V_{feedback}}{\Delta V} \qquad (2.25)$$

- The statism k_{stat} or the relative balancing error, equal to the ΔV is always constant from the input value

$$k_{stat} = \frac{\Delta V}{V_x} = \frac{\Delta V}{V_{feedback} + \Delta V} = \frac{\Delta V}{\Delta V(1 + k_{sub}k_{dc}k_{ic})} = \frac{1}{1 + k_{sub}k_{dc}k_{ic}} \qquad (2.26)$$

● The relative balancing depth:

$$k_{\text{depth}} = \frac{\Delta V_{\text{feedback}}}{V_x} = 1 - k_{\text{stat}} \qquad (2.27)$$

Taking into account received characteristics

$$f_x = \frac{V_x}{k_{ic}}(1 - k_{\text{stat}}) \approx \frac{V_x}{k_{ic}}, \text{ if } k_{\text{stat}} \ll 1 \qquad (2.28)$$

$$\delta k_{\text{VFC}} = \delta k_{dc} k_{\text{stat}} + \delta k_{\text{sub}} k_{\text{stat}} - \delta k_{ic}(1 - k_{\text{stat}}) \qquad (2.29)$$

Neglecting the error δk_{sub} owing to its smallness we have:

$$\delta k_{\text{VFC}} = \delta k_{dc} k_{\text{stat}} - \delta k_{ic}(1 - k_{\text{stat}}) \qquad (2.30)$$

In order that the conversion error including the δk_{stat} is insignificant, it is necessary that $k_{\text{stat}} \ll 1$. The limit of k_{stat} reduction is the loss of the system stability at which there is an abnormal mode of the $V_x \rightarrow f_x$ conversion. The optimum values k_{stat} and the loopback amplification are the following:

$$k_{\text{stat\_optimal}} = \delta k_{ic}/(3-5)\delta k_{dc}; \quad (k_{dc} \cdot k_{ic})_{\text{optimal}} = \frac{(3 \dots 5)\delta k_{dc}}{\delta k_{ic}} \qquad (2.31)$$

It is necessary to note that apart from the high speed, the absence of the conversion error caused by the incomplete balancing is another important advantage of such a VFC. Since this error has a regular character it may be taken into account by the VFC graduation. It is possible because k_{stat} does not depend on the measuring voltage V_x, therefore it may be easily corrected. The conversion error because of ΔV will take place only in the case of the statism inconstancy. However, it can be reduced by calibration of its regular, i.e. constant component.

The distinctive feature of the VFC with switching of the integration direction is the use of the capacitor recharging for conversion of alternating voltages. Such a VFC consists of the control charging devices, switches, an integrating capacitor, a comparison device of the reference voltage and the capacitor voltage and a control trigger (Figure 2.10).

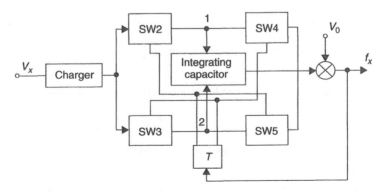

Figure 2.10   VFC with switching of integration direction

The VFC works as follows. The voltage V_x is applied to the control charging device's input, the output current of which is the charging current of the integrating capacitor. If, at the first moment, the switches SW2, SW5 are closed and SW3, SW4 opened, the circuit I charges the integrating capacitor. Therefore

$$V_1 = \frac{1}{C} \int_0^{T_x} I_{\text{charge}} \, dt, \; V_1 = V_0 \tag{2.32}$$

At the moment of equality $V_x = V_0$, the pulse, which switches the control trigger, is formed. Switches SW2, SW5 are opened, and SW3, SW4 closed. The new circuit of the capacitor charge is formed under the current influence, the integrating capacitor is recharged and the potential in point 2 is changed from $-V_0$ up to $+V_0$. At the moment of equality $V_2 = V_0$ the process begins again. We shall notice, that in both cases, the potential in point 1 (in the first case) and point 2 (in the second case) grows linearly up to the reference potential with a speed proportional to the value of the input voltage V_x. Thus, there are oscillations in the circuit. We will have pulses, whose frequency depends on the value of the input voltage V_x on the comparison device's output.

The charger device supports the current I_{charge} constant at the integrating capacitor charging at constant V_x. The voltage on it depends linearly on time. From Equation (2.32) we find the time T_x of the capacitor recharging from $-V_0$ up to $+V_0$ and the basic equation for the $V_x \rightarrow f_x$ conversion:

$$T_x = \frac{C \cdot V}{I_{\text{charge}}} = \frac{2C \cdot V_0}{I_{\text{charge}}} = \frac{2C \cdot V_0}{k \cdot V_x}; \quad f_x = \frac{1}{T_x} = \frac{k \cdot V_x}{2C \cdot V_0}, \tag{2.33}$$

where k is the conversion factor of a charger device.

It is obvious, that in order to have the linear dependence of the VFC's output frequency on the value V_x it is necessary for the integrating capacitor to have stable parameters, constant reference voltage, and a charger device with a straight-line characteristic and large stabilization of the current I_{charge}. The experimental results are the following $V_x = (0.1–1.5)$ V; $f_x = (500–20\,000)$ Hz; the nonlinearity error is 0.3%.

The best VFCs have almost as good performance and provide unique characteristics when used as analog-to-digital converters, but are relatively slow. A lot of VFC converters exhibit some nonlinearities in comparison with dual-slope converters.

New architectures of integrated VFC converters with the charge-balance, the inversed VFC, the synchronized VFC and a relatively fast synchronous version of a VFC with the charge balance were recently described in [82]. Their versatility, excellent resolution, very wide dynamic range and an output signal that is easy to transmit often provide attributes unattainable with other converter types. Since the analog quantity is represented as a frequency of a serial data stream, it is easily handled in extensive industrial multichannel systems. Virtually unlimited voltage isolation (tens of thousands of volts, or even more) can be accomplished without loss in accuracy using low-cost optocouplers or fibre optic links. At the other end, the digital signal can be either digitally processed or reconstructed to analog, which is useful in many, for example, medical, applications. A VFC possesses true integrating input and features the best noise immunity, much better than a dual-slope converter. It is especially important in industrial measurement and data acquisition systems.

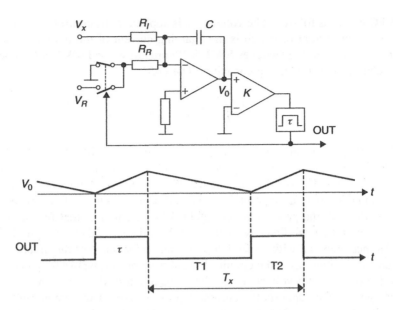

Figure 2.11 Voltage-to-frequency converter with charge balance and its time diagrams (Reproduced by permission of Silesian Technical University)

A voltage-to-frequency converter is actually a kind of precise relaxation oscillator. Nearly all modern VFCs use a well-proven charge-balance technique to control the output frequency, as shown in Figure 2.11.

The integrator produces a two-part linear ramp: the first part (T1) is a function of the input voltage, the second (T2) is dependent on both the input voltage and the constant reference current. The oscillation process forces a long-term balance of the charge (or time-averaged current) between the input current and the reference current:

$$V_x \cdot \frac{T_1 + T_2}{R_I C} + V_R \cdot \frac{T_2}{R_R C} = 0 \qquad (2.34)$$

If

$$\frac{1}{T_2} = F_{max} \text{ and } \frac{1}{T_1 + T_2} = F_x \qquad (2.35)$$

then

$$-\frac{V_x}{V_R} = \frac{R_I}{R_R} \cdot \frac{F_x}{F_{max}}. \qquad (2.36)$$

If both resistors R_1 and R_R are identical, the result of conversion is independent of drifts of the integrating components. The converter's integrating input properties make it ideal for high noise industrial environments and guarantees excellent noise immunity by smoothing the noise effects. Every change in the input signal always affects the output in the current or the next conversion cycle so averaging consecutive results improves overall accuracy and noise attenuation.

The accuracy of a VFC depends mainly on the period T2 of the one time. In precision, synchronous VFCs in this critical period T2 are accurately derived from

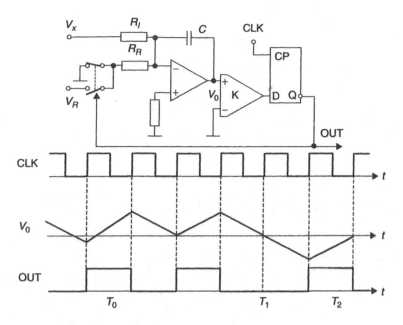

Figure 2.12   Synchronized voltage-to-frequency converter and its time diagram (Reproduced by permission of Silesian Technical University)

an external, crystal controlled oscillator (Figure 2.12). When the integrator negative-going ramp reaches zero, the integration of the input signal will continue, awaiting the clock pulse. Output pulses align with clock pulses, which cause the instantaneous converter output frequency to be a subharmonic of the clock frequency. The time intervals between pulses become not equal, but the average frequency is still an accurate analog of the input voltage. The output may be presented as a constant frequency, which is locked to a submultiple of the clock frequency, with occasional extra pulses or missing pulses.

All voltage-to-frequency converters exhibit some nonlinearities in comparison with dual-slope converters. The linearity performance decreases and the gain error increases with the full-scale operating frequency. It is obvious that the integrator in a dual-slope converter changes the direction of the output ramp about 10, 20 or 50 times per second, in a VFC a few thousand or even more. Delays in the integrator and analog switches generate errors coupled with every transition and are proportional to the output frequency.

When the system utilizes a VFC, an appropriate frequency measurement technique must be chosen to meet the conversion accuracy and speed requirements. The commonly used method for converting the output of a VFC to a numerical quantity is to accumulate the output pulses in a gated counter. The quantization error can be made arbitrarily small by counting with long gate times. However, many applications require relatively fast conversions (short gate periods) with high resolution.

The inversed VFC integrating analog-to-digital converter seems to be quite attractive. In this converter, the input signal is continuously integrated, as in the VFC and Elbert's converters, providing excellent noise immunity. Working very much like the dual-slope ADC, it features similar, or even a little better, properties. The circuit, presented in

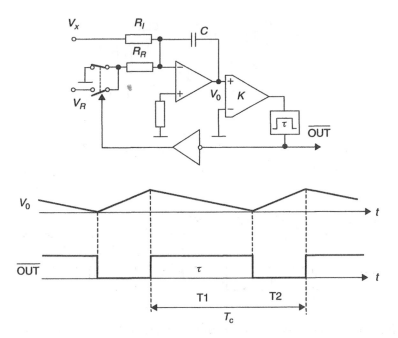

Figure 2.13   Integrating inversed VFC and its time diagrams (Reproduced by permission of Silesian Technical University)

Figure 2.13 differs slightly from the dual-slope architecture–one of the switches is removed. Actually, it's very similar to the V/F converter configuration.

As in other integrating converters, the integrator produces a two-part linear ramp: the first time period (T1) is constant, the second (T2) is dependent on the input voltage and the constant reference voltage of the opposite polarity. During the time T1 only the input signal V_x is integrated and the integrator's output ramps linearly to the level proportional to the input voltage. In the second phase of conversion the reference voltage V_R is also switched to the integrator input, which processes to integrate in an opposite direction. The length of the time T2 the integrator output requires to get back to zero depends on the input voltage level and is equal to zero for the grounded input, but goes to infinity for I_X close to I_R. When the ramp reaches zero, the comparator activates the one-shot for the period T1. This disconnects the reference voltage from the integrator input again. Integrator slopes are both established by the same integrating capacitor and the two resistors' ratio, so these parameters need to be stable over one conversion time. The integrator's swings during both phases of conversion are equal and the capacitor charge is balanced:

$$V_x \cdot \frac{T_1 + T_2}{R_I \cdot C} + V_R \cdot \frac{T_2}{R_R \cdot C} = 0 \text{ and } -\frac{V_x}{V_R} = \frac{R_I}{R_R} \cdot \frac{F_x}{F_{\max}} \qquad (2.37)$$

These equations are exactly the same as Equation (2.36), describing the voltage-to-frequency converter; however, both converters behave quite differently–the time T2, not T1, is constant in a VFC. According to these similarities and differences the name of this converter is the inversed VFC, or the IVFC.

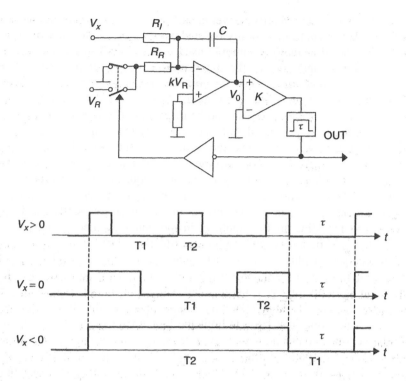

Figure 2.14 Modified inversed VFC and its time diagrams (Reproduced by permission of Silesian Technical University)

The primary circuit can be modified. Figure 2.14 shows one of the possibilities.

A voltage divider with the attenuation $1/k$ connected to the reference source offsets the non-inverting input of the operational amplifier. An analog switch connects the reference resistor R_R to the ground during the phase T1 and to the voltage V_R during T2, so the inverting input of the operational amplifier always sees the constant resistance. The following equation describes the charge balance during the conversion period T1 + T2:

$$\frac{V_x - k \cdot V_R}{R_I} \cdot (T_1 + T_2) + \frac{(1 - k) \cdot V_R}{R_R} T_2 - k \cdot \frac{V_R}{R_R} \cdot T_1 = 0 \qquad (2.38)$$

$$\frac{V_x}{V_R} = k \cdot \left(1 + \frac{R_I}{R_R}\right) - \frac{R_I}{R_R} \cdot \frac{T_2}{T_1 + T_2} \qquad (2.39)$$

For $R_I = R_R$ and $k = 0.25$

$$\frac{V_x}{V_R} = 0.5 - \frac{T_2}{T_1 + T_2} \qquad (2.40)$$

In this case, the circuit converts the bipolar input voltages. If the input ranges is limited to $\pm 0.25 V_R$, T2 changes from 0.33T1 to 3T1; for $V_x = 0$, T2 = T1.

This converter shares many properties of other integrating dual-slope ADC, AFC and Elbert's converters. Its integrating input characteristics provide excellent noise immunity in industrial environments. The time-related, two-phase output signal is almost as easy to transmit as in Elbert's converter and also contains information about the input polarity. This signal can be easily isolated using opto-couplers or fibre-optic lines and can be interfaced to many commonly used microcontrollers and microprocessors through the counter input port or the timer/counter peripheral ICs. The accuracy is comparable with the dual-slope, while the noise immunity and the ease of transmission to VFC or Elbert's converters. For determining the result of the conversion, times T1 and T2, or only the duty factor D_F should be measured. The transmitted digital signal can be not only digitally processed, but also reconstructed to the analog one.

A growing proportion of complex data acquisition and control systems today are being controlled by microprocessors or other programmable digital devices. As mentioned, the use of voltage-to-frequency converters requires an appropriate frequency measurement technique. Although it is not a very fast conversion method, a VFC system conversion speed can be optimized. In the most frequently used approach output pulses of the VFC are accumulated in the gated counter during the constant time. Since the gate period is not synchronized to the VFC output pulses, there is a potential inaccuracy of plus or minus one count–the resolution is related to the gate period and to the full-scale frequency of the VFC. The quantization error can be made small by counting with long gate periods and using high output frequencies, but the VFC linearity degrades at high operating frequencies and this limits the accuracy. On the other hand, long gate periods limit the conversion speed. Many applications require relatively fast conversions with good resolution and accuracy. The ratiometric counting technique partially eliminates this tradeoff; by counting n counts of a high speed clock (independent of the VFC clock) which occur during an exact integer N counts of the VFC, an accurate ratio of the unknown frequency F_x to the reference frequency so the n count is a large number. The one-count error causes a small effect on the conversion result. The two counts (n and N) are divided in a host computer or a microcontroller, giving the final result. Since the synchronized gate waits complete cycles of the VFC to achieve an exact count, a very low frequency would cause the gate period to be excessively long. This can be eliminated by offsetting the zero-input frequency to be constant and stable, and time intervals between pulses are exactly equal. Unfortunately, in precision, synchronous voltage-to-frequency converters (as in Figure 2.12), these conditions do not occur.

Figure 2.15 shows a modified, precision and relatively fast synchronous version of a VFC with the charge balance.

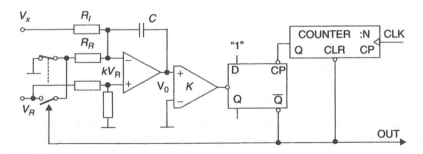

Figure 2.15 Modified 'fast' version of a VFC (Reproduced by permission of Silesian Technical University)

Table 2.1 Summarized typical parameters of different integrating converters (Reproduced by permission of Silesian Technical University)

Converter parameter	Dual-slope	VFC	Elbert's	IVFC
T1	constant	variable	variable	constant
T2	variable	constant	variable	variable
V_x integration	T1	T1 + T2	T1 + T2	T1 + T2
Noise immunity	good	excellent	excellent	excellent
Accuracy	high	good	excellent	high
Integrator swing	variable	nearly constant	constant	variable
V_x/V_R	T2/T1	T2/(T1 + T2)	(T1 − T2)/(T1 + T2)	T2/(T1 + T2)

In this converter the disintegrating period T2 is created by the counter, which counts the high speed clock pulses. Because the time T2 is constant and stable, for determining the conversion result the time interval between two pulses, or only time T1 can be measured. It varies within one clock period limit and the resolution increases as the clock frequency and the counter capacity N increase. Only one output pulse period measurement is sufficient which means that for the given conversion speed, the VFC can operate at very low frequencies where its linearity is excellent, or, for the given operating frequencies, the conversion time can be considerably shorter. To prevent too long output pulses intervals, the converter should be offset for example by a voltage divider with the $1/k$ attenuation. The accuracy and the conversion speed of the converter are comparable to those of the dual-slope converter, while the noise immunity is much better. The output signal is very easy to transmit over long optical or electrical lines.

Equations describing this converter are exactly similar to Equations (2.38–2.40). Similar to the circuit in Figure 2.14, this converter accepts bipolar input voltages. If the input range is limited to $\pm0.25V_R$, T1 changes from 0.33T2 up to 3T2; for $V_x = 0$, T1 = T2.

The summarized typical parameters of different integrating converters are shown in Table 2.1. The conversion cycles consist of two phases T1 and T2.

2.2 Capacitance-to-Period (or Duty-Cycle) Converters

Capacitive sensors have received a wide distribution, for example, as primary converters of humidity, pressure, movement and many other non-electrical parameters. Among many analog-to-digital conversion techniques, a simple solution consists of using variable oscillators coupled with counters. A capacitance sensor can be used in a simple oscillator circuit to provide a frequency output that is inversely proportional to a measurand.

With capacitive sensors there are problems of transducer creation, providing the conversion invariance to additional parameters of the sensor's electrical equivalent circuit and, to the leakage resistance, $R_{leakage}$. Difficulties of this problem are increased by the fact that the leakage resistance is a complex function on the measurand as well as on the set of influencing factors. It results in the essential error, taking into account of which is extremely difficult. One of possibilities for converter design, providing

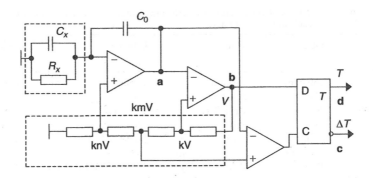

Figure 2.16 Capacitance-to-voltage-to-duty-cycle converter

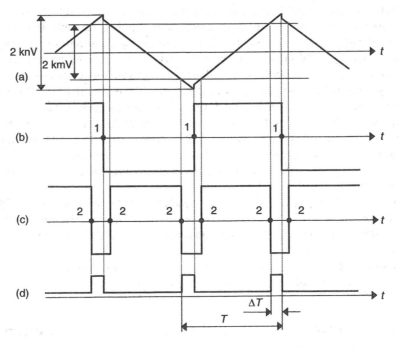

Figure 2.17 Time diagrams of output voltages (a–operation amplifier; b and c–comparators; d–trigger)

invariance to the leakage resistance is the use of parallel output signal processing of the sensor [83].

The circuit of such a converter and its time diagrams are shown in Figure 2.16 and Figure 2.17 accordingly. The capacitive sensor presented by the two-element parallel equivalent circuit (R_x, C_x) is connected to the operational amplifier's input, with the capacitive negative feedback through the reference capacitor C_0. At the operational amplifier's output signal processing, carrying the information about values of leakage capacitance and resistance, two identical comparators are used. Its levels of comparison, formed from the output voltage $\pm V$ of the first comparator with the help of the voltage divider, are various and equal accordingly to $\pm kV$ and $\pm kmV$ ($k, m =$ const). The input voltage of the operational amplifier is equal to $\pm knV$ ($n =$ const) and is formed from the same voltage.

The converter is working in an unstable mode. From the analysis of Figure 2.17 it follows that for the moments 1 of the comparator operation the following equality is valid

$$knV \cdot \left(1 + \frac{C_x}{C_0} + \frac{T}{2 \cdot C_0 \cdot R_{\text{leakage}}}\right) = k \cdot V. \tag{2.41}$$

Its output voltage changes a sign during each half-cycle of oscillation at the moment of 2 (Figure 2.17) equality of voltages on the second comparator's input. The time interval ΔT between comparators' operation is determined from the ratio

$$kmV + knV\frac{\Delta T}{2 \cdot C_0 \cdot R_{\text{leakage}}} = kV. \tag{2.42}$$

Solving Equations (2.41) and (2.42) together we have:

$$\frac{T}{2 \cdot \Delta T} = \frac{n}{1 - m} \cdot \left(\frac{1 - n}{n} - \frac{C_x}{C_0}\right). \tag{2.43}$$

If we take into account that $C_x = C_{\text{nom}} \pm \Delta C$, where ΔC is the change of sensor's capacitance under measurand influence and also $C_0 = C_{\text{nom}}$, we have:

$$\frac{T}{2 \cdot \Delta T} = e_1 \cdot \left(e_2 \pm \frac{\Delta C}{C_{\text{nom}}}\right), \tag{2.44}$$

where $e_1 = n/(1 - m)$; $e_2 = (1 - 2n)/n$ are the dimensionless constant factors, whose stability is determined by the stability of the divider's division factors. At $m = 2n$ and $\Delta C/C_{\text{nom}} = 0$, $T/2\Delta T = 1$.

Intervals ΔT and T are formed by the trigger. The measurement of these intervals and calculation of its ratio is realized by the microcontroller.

Thus, the described circuit provides the full independence of conversion results on values R_{leackage} at rather simple circuit realization. The main error of the transformation does not exceed 0.2%, the conversion time is not more than 10^{-2} s.

The capacitance-to-period converter for pressure sensors is described in the paper [84]. This transducer is based on a capacitance ratio to the frequency ratio conversion in order to procure a high level of self-compensation for temperature drifts and nonlinearity. The equation of the period is:

$$T = \tau_d + K \cdot (C + C_s), \tag{2.45}$$

where

$$K = k \div \left[1 - \left(\frac{I_f}{I_0}\right)^2\right] \tag{2.46}$$

and I_f is the leakage current; C_s is the stray capacitance; $\tau_d = \tau_1 + \tau_2 \approx 4\Delta t$ is the offset. As the universal period meter and different ceramic capacitors have evaluated the converter response with a good resolution (0.01% of the measured value).

The system, which combines passive telemetry with a pressure measuring system, is based on a capacitive type the pressure sensor and ASIC chip, which converts capacitance variations into frequency, is described in [85,86]. The output frequency (up to 160 kHz at 0 mmHg) can be calculated as:

$$f = \frac{k\dfrac{W_0}{L_0}(V_{\text{bias}} - V_{TN})}{2C_x(\sqrt{n} - 1)}.$$

(2.47)

This equation implies that the output frequency is independent of supply voltage and dependent on temperature through the mobility term in k and the threshold voltage V_t. The bias voltage V_{bias} is chosen to be equal to the bandgap reference voltage produced from the previous stage and is thus considered independent of voltage and temperature variations.

The pressure measuring subsystem consists of a capacitive type pressure sensor and a C/F converter integrated circuit to immediately convert capacitance into the electrical signal. The C/F chip includes the internal voltage regulation and a current-mode comparator, which improves the stability of the output frequency. The C/F dice takes up 1.44 mm^2.

The conversion of non-electrical parameters to frequency can also be based on the use of inductive sensors according to one of the following ways: (1) according to the sensor's inductance the frequency of the LC-generator is determined; (2) the sensor is connected to the bridge circuits, balanced by the measurement of the voltage frequency applied to the bridge; (3) the inductance sensor is connected to the selective RL-circuit of the self-oscillator. Today microelectronics and microsystems give us the possibility to realize the smart sensor microsystem including the integrated electrodeposited flat coil and the interface circuit on the same substrate.

Summary

All the described VFC converters share many properties. They have a similar configuration to the integrator, analog switches and the comparator. Many of them can be integrated to smart sensors in order to produce further conversion in the quasi-digital domain instead of the analog domain.

As considered earlier, modern CAD tools contain the microcontroller core and peripheral devices as well as voltage-to-frequency converters. So, for example, the Mentor Graphic CAD tool includes different kinds of VFCs such as AD537/650/652, the CAD tool from Protel includes many library cells of different Burr–Brown's VFCs. The realization of other converters, for example, the capacitance or the resistance-to-frequency are not technological problems for single-chip implementation. All this makes the transition from the analog signal domain to the quasi-digital (frequency-time) domain easy enough.

3

DATA ACQUISITION METHODS FOR MULTICHANNEL SENSOR SYSTEMS

Technological and manufacturing processes are sources of the initial data for multi-channel sensor systems. The data are received by a control system as random processes of parameter variations and in the form of random events.

Multichannel data acquisition systems are intended for the transformation of the initial parameters of processes and events (output signals from one or several sensors and transducers) into equivalent digital signals, suitable for further processing, transferring and input into a central computer, which controls the channeling data acquisition and forms data arrays for further display or its use in control systems. Modern data acquisition systems are able to handle practically all physical and chemical quantities, due to the wide variety of frequency–time-domain sensors and transducers.

Methods of data acquisition depend on solved tasks of control and measurement and directly influence the structure and functionalities of multichannel data acquisition systems.

A modern measurement and control sensor system could be set up in different ways. A central computer is connected to a number of input (sensors) and output (actuators) devices. In such systems, many frequency–time-domain sensors pick up information about process-related measurands.

Two traditional methods of data acquisition are widely used in modern automatic control and measuring systems. These are:

- Methods using time-division channelling, based on multiplexing sensors, i.e. on time-shared data acquisition from each sensor.

- Methods using space-division channelling, based on simultaneous data acquisition from all the sensors.

In both cases, the constancy of data sources, i.e. access to information at any time dependent on solved control and measuring tasks, is used.

3.1 Data Acquisition Method with Time-Division Channelling

The most frequently used configuration of a data acquisition system with time-division channelling (multiplexing) is shown in Figure 3.1.

In this system outputs of the frequency sensors $f_1, f_2, \ldots f_n$ are connected to the frequency-to-code converter F/# in turn with the help of the digital multiplexer MX, which is controlled by the microcontroller μK. The frequency-to-code converter converts the frequency f_x to a binary code, for example, according to the direct counting method by counting $T_x = 1/f_x$ periods during the gate time T_q (quantization time) or according to the indirect counting method by counting impulses of the high reference frequency f_0 during one T_x or nT_x periods. In modern data acquisition systems for frequency output sensors, the frequency-to-code conversion can be realized directly by microcontrollers without any additional hardware.

After the frequency-to-code conversion, the binary code enters the microcontroller to form data arrays. If necessary, additional signal processing, for example, linearization, unification, scaling, etc. can be realized in the microcontroller or DSP. The data is then input into a central computer (through one of the system buses, I/O ports or with the help of direct memory access (DMA)) for further processing, display or use in the control system.

Sensor polling can be cyclic synchronous as well as software-controlled asynchronous—the microcontroller chooses the required sensors depending on the task. In data acquisition systems of data with frequency conversion at cyclic polling with constant time intervals between identical operations the cycle polling time τ_0 can be calculated according to the following equation:

$$\tau_0 = n \cdot (T_q + \tau_{\text{delay1}} + \tau_{\text{delay2}}), \tag{3.1}$$

where:

— T_q is the quantization time in the frequency-to-code converter.

— τ_{delay1} is the time delay between the ending of the conversion in the previous sensor and the command to poll the next sensor.

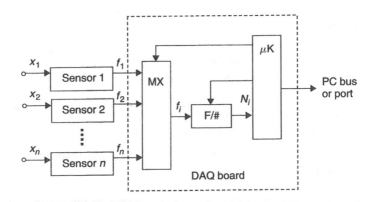

Figure 3.1 A Data acquisition system with time-division channelling

— τ_{delay2} is the time delay of the frequency conversion starting after the sensor connection.

— n is the number of sensors in the multichannel data acquisition system.

Values of the time intervals and delays are determined by methods of the measurement of informative parameters, methods of separation and electronic components used. At the preset values T_q, τ_{delay1}, τ_{delay2} and the cycle polling time τ_0 can be changed only by changing n.

The constant sequence of sensor polling and the cyclicity, controlled by the microcontroller, is the reason for information losses. In order to reduce these losses, it is necessary to either increase the frequency of sensor polling or to use other technological and algorithmic measures.

The multichannel device for data acquisition based on the method of accelerated polling for period output sensors is shown in Figure 3.2. It simultaneously forms the period duration T_x and the sensor's number in its output [87]. The device contains some RS-triggers (according to the number of channels in the system) with multiport elements AND, which works similar to a frequency switch, and a coder CD for conversion of the input positional code to the binary code of channel's number. In this device, signals from periodic output sensors arrive to synchronous inputs of D-triggers of register 1. The coder's output and the output of the first multiport element AND 4 form the

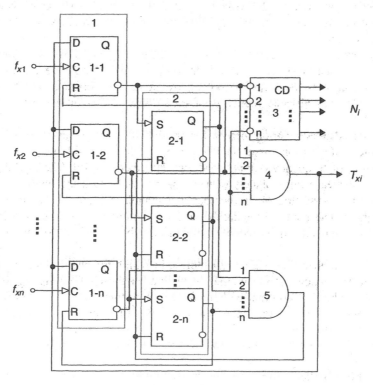

Figure 3.2 Device for data acquisition based on the method of accelerated polling for periodic output sensors

device's output. The synchronous inputs of D-triggers and the S-input of RS-triggers are dynamic inputs which toggle from a positive logic swing, other inputs are potential.

The device works as follows. At the beginning of each cycle all triggers of both registers are in the logical state '0'. In some moment of time the D-trigger connected to the ith channel is toggled to '1' by the positive logic swing. From the inverse output of this trigger, logic '0' arrives at the i-th input of the multiport logic element AND 4, the S-input of RS-trigger 2 and the ith input of the coder where it will be converted to the binary code of channel's number. The logic level '0' of multiport logic element AND 4 forbids toggles of all other D-triggers of the register '1' until the next positive logic swing in this channel. At this logical level the D-trigger in the ith channel is toggled to '0'. The RS-trigger in the same channel is also toggled. This trigger locks the D-trigger at R-input by the positive logic swing on the inverse output up to the cycle ending. Thus, the logical '0' output of the multiport element AND 4 and the binary code of the chosen channel's number on the coder's output remain constant during the pulse propagation in this channel.

The next toggled trigger will be the D-trigger of that channel, the C-input of which will then act as the logic swing before all others and all described actions will repeat again. After the toggle '1' of the RS-triggers of all n channels, the second multiport element AND 5 will toggle and reset the register based on RS-triggers to the '0' level.

This device has extended functionalities, due to the formation of the time T_{xi} of the multichannel period-to-code converter. The polling of all measuring channels is realized during one cycle. Moreover it is carried out with high speed because the next measuring channel is chosen according to the nearest logic swing from all the nonselected channels.

A similar method of accelerated polling for quasi-digital sensors was also used in multichannel digital frequency selector design [88]. The distinctive feature of these frequency selectors is the increased polling frequency for channels with higher input frequencies that has allowed reduction of the dynamic error of measurement in this channel.

The original data acquisition method for multichannel frequency-to-code converters which reduce information losses in multichannel data acquisition systems by time-shared polling of frequency output sensors is described in [89]. The method consists of a choice of measuring channels with the nearest pulses on the phase of converted frequencies; the separation of the number of channels, which can be simultaneously served by the counter; storing the number of these channels in memory; the connection of the chosen channels to the counter's input and the frequency-to-code conversion in them.

Further, the numbers of channels are stored, in which repeatedly there was no code of converted frequencies at the moment of arrival of the challenging signal from the outside, and applied extraordinarily their connection to the counter's output with arrival to these channels of the nearest pulses on time. Then the number of channels in which measurement of time intervals is supposed will be stored, and uses its extraordinary connection to the counter's input when the start pulses arrive, so beginning the time measurement.

This method of data acquisition allows minimum information losses because of nonsyncronism and nonphasing converted and polling frequencies. This arises because of the channels' sequence of connection to the counter's input results in the increased probability of the next pulses of converted frequencies occurring at once after the ending of the count in previous channels, as the sample is carried out from all sets of converter channels.

Thus, channel choice is determined as a result of a comparison of information in all the channels by revealing the first pulse of one of channels from which the control pulses are forming for processing units.

Data acquisition with time-division channelling allows inexpensive multichannel data acquisition systems for quasi-digital sensors.

3.2 Data Acquisition Method with Space-Division Channelling

The most frequently used configuration of a data acquisition system with space-division channelling is shown in Figure 3.3.

In such a system instead of one frequency-to-code converter and an n-channel multiplexer, n frequency-to-code converters (according to the number of channels) and a microprocessor system with n inputs are used. That is, for simultaneous measurement of several frequencies, there is an independent channel for the frequency-to-code converter. The microprocessor simultaneously starts all converters (continuous periodic synchronously or software-controlled asynchronously), and at the end of the measurement processes reads results. It is possible to make some realizations of the frequency-to-code converter — the microprocessor interface: by polling, interrupt or DMA. Due to n independently working measuring channels (the sensor — the frequency-to-code converter — the microprocessor's input), n conversions are realized simultaneously. It increases the system productivity and speed by n times which is determined by the time $T_q + t_{\text{readout}}$. However, this important advantage — the reduction of measuring information losses — is achieved by additional hardware and cost. All this limits the number of channels in a data acquisition system or requires the use of special technical measures, for example, the separation of some element from the converter, which can be common for all frequency-to-code converters, etc.

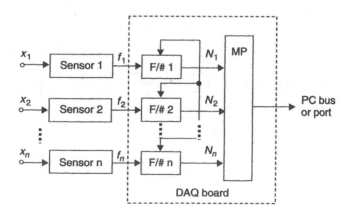

Figure 3.3 Data acquisition system with space-division channelling

One such solution is used in a device for multichannel frequency conversion [90]. Simultaneously with hardware reduction and parallel measurement of all sensor frequencies, accuracy is increased without increasing time of measurement, due to the elimination of the influence of sensor frequency fluctuations. The circuit of such a device for simultaneous conversion of frequencies f_1, f_2, ... f_n, is shown in Figure 3.4. Each channel contains a frequency divider, a switch and a buffer register. The counter of the reference frequency f_0, and the register are common for all measuring channels.

Let's consider how this device works in detail. For measurement of any input frequency the microprocessor generates a command, which regulates the divider overflow pulse propagation through switches to the buffer registers. The first overflow pulse from the divider starts the 'calibration'. This pulse puts the logic '1' in the appropriate register's bit and copies the counter's content into the buffer register. Simultaneously, the overflow pulse propagation for writing is forbidden until the new command from the microcontroller. During the frequency measurement, information from the buffer register is readout by the microprocessor at the necessary moment of time. After the reading this register will be nulled. The information from the register about divider overflow and the counter of reference frequency is readout by the microprocessor periodically with a greater frequency, than the frequency of overflow pulses. The overflow pulse frequency is determined by the counter capacity and the frequency of pulses to its input. It excludes pulse losses. At the end of the measurement cycle of frequency, a command similar to the command at the beginning of the cycle is generated by switches. The next divider overflow pulse after this command determines the end of a measurement cycle. It writes the information from the reference frequency counter into the buffer register for further readout by the microprocessor.

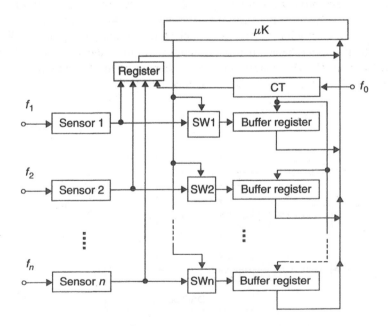

Figure 3.4   Device for simultaneous multichannel frequency conversion

Thus, the following information enters the microprocessor during frequency conversion:

— K_1 is the number of divider overflows.

— K_0 is the number of reference frequency counter overflows.

— n_1 is the number written into the reference frequency counter at the beginning of the measurement cycle.

— n_2 is the number written into the reference frequency counter at the end of the measurement cycle.

The converted frequency can be calculated according to the following equation:

$$f_1 = \frac{N_1 \cdot K_1}{N_0 \cdot K_0 - n_1 + n_2} \cdot f_0, \tag{3.2}$$

where f_0 is the reference frequency; f_1 is the input frequency (measurand); N_1 is the divider capacity; N_0 is the reference frequency counter capacity.

The device for frequency measurements receives some measurement results during one cycle. The frequency for each measurement is determined by these results and the received results are averaged. It reduces the measurement error from pulses fluctuations. Thus, the measurement cycle duration is constant.

During simultaneous measurement of several frequencies the dividers' overflow pulses enter the appropriate bits of the register. The reading of information from buffer registers is separated in time. Hence, measuring frequency channels do not influence each other.

The method of data acquisition for quasi-digital sensors with space-division channelling is becoming increasingly attractive due to the small cost of frequency-to-code converters. A separate frequency-to-code converter for each channel allows a much higher polling frequency for each channel.

3.3 Smart Sensor Architectures and Data Acquisition

Depending on the smart sensor architecture, various data acquisition schemes are possible. Let us consider some of them. In the smart sensor architecture shown in Figure 3.5, the analog output of the sensor element S is at first amplified and corrected for offset, non-linearity, etc. Then the voltage-to-frequency conversion takes place. The frequency–time-domain signal (frequency, period, time interval, duty-cycle, etc.) is converted into code. The format of the frequency-to-code converter is such that the signal is transferred to the bus system at the command of the bus controller.

In the second example (Figure 3.6) sensor elements form a sensor array. A single multiplexing circuit feeds signal-conditioned signals from the sensing elements one after the other into a single frequency-to-code converter and from here, the signals are transferred to the bus. Such a sensor array or a multiparameter sensor can, for example, measure different variables such as temperature, pressure, humidity, etc. at a certain location.

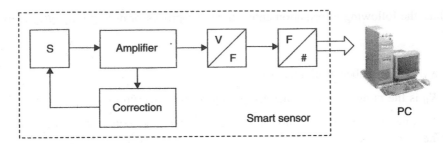

Figure 3.5   Smart sensor architecture with preliminary correction in analog domain and further conversion into frequency–time-domain signal

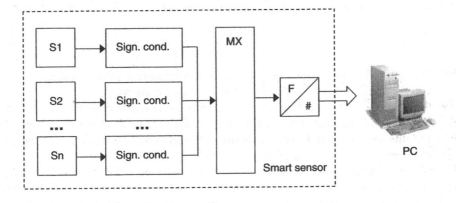

Figure 3.6   Architecture of a smart sensor array

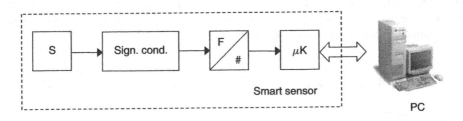

Figure 3.7   Smart sensor architecture with microcontroller

In the third example, a sensor element is connected via a frequency-to-code converter to a microcontroller (Figure 3.7).

The microcontroller can store the sensor's characteristic data in its internal ROM and, based on this information and the sensor signal, the microcontroller transfers the corrected signal to the bus. A very useful feature of such smart sensor architecture is that the microcontroller also permits the central computer to send data back to the sensor, which can be used to change the measuring range, to exert a recalibration or to adjust the offset.

As mentioned earlier, the microcontroller can itself realize the frequency-to-code conversion. In this case, the smart sensor architecture becomes simpler. Such an architecture is used to achieve smart sensor self-adaptation when depending on measuring conditions or measurands, the parameters of the method for frequency-to-code conversion will vary, for example, conversion time or accuracy of measurements.

In such a subsystem with the bus architecture, each sensor or a group of sensors can also contain a circuit, which can recognize addresses, this means the circuit can detect when communication between the sensor and the central computer is desired.

For many years, most of the described components of a bus system were separated and had their own housing. However, more recent developments have enabled the components indicated by the dashed lines in Figures 3.5–3.7 to be integrated into a single chip by the implementation of standard microelectronics library cells.

These typical smart sensor architectures are used for the creation of digital output smart sensors. Data transmission in the digital form excludes the inphase interference and the voltaic coupling of the sensor's output with the computer can be provided if necessary. Instead of buses, the binary encoded information can be transmitted into a parallel or a serial port. When parallel data transmission is feasible, the pulses representing the information arrive simultaneously at the central computer input making fast data transfer possible. However, in most cases the distance between the sensor and the computer is too far to permit parallel data transfer and so serial data transmission is required. The main advantage of such an approach is that it is not necessary to use any additional computer boards and specialized software drivers. All connections are external and drivers are standard. External connection provides an additional coupling in comparison with the usual data acquisition system with the bus connection. As the cost of microcontrollers continues to decrease, using a microcontroller at each measurement location will become affordable. In the future, many different signal-processing circuits can be integrated into the sensor chip. This approach is the next step to much wider distribution of intelligence. From Figures 3.5–3.7 the most important elements of a bus-oriented data acquisition system can be deduced. It is expected that starting with large data acquisition systems, analogue data transfer will gradually be replaced by digital systems.

DAQ boards for frequency–time-domain sensors are produced by many companies. They carry out data acquisition and data processing. Technical performances of some of them are shown in Table 3.1.

The maximum frequency and the number of channels meet modern requirements. But the accuracy is not perfect in order to use some of these DAQ boards with the modern frequency (period, duty-cycle, time interval) output sensors. Besides, it is desirable to have an opportunity to change both accuracy and time of the measurement directly during data acquisition.

3.4 Main Errors of Multichannel Data Acquisition Systems

The first measuring conversion in a channel of the data acquisition system is realised by a frequency (period) output sensor with information ability

$$I_D = \lg_2 \left(\frac{100}{2\gamma_S} + 1 \right), \text{ bit} \tag{3.3}$$

Table 3.1   Technical performances of DAQ boards for frequency–time parameters of signals

Type	Maximum source frequency, MHz	Number of channels	Base clock accuracy, %
	Timing I/O boards from National Instruments		
PC-TIO-10	7	10	0.01
NI 660X	20–80 (60–125)*	4–8	0.005
	PCI data acquisition boards from Keithley		
KPCI-310XX	5 (20)**	3–4 (8)**	N/a
	Frequency-input card from IOTECH		
DBK7	0.95	4	Accuracy 0.1 %
	Digital I/O and timing card from Meilhaus Electronic		
ME-1400A/B	10	3–6	0.01

*with prescalling; **for KPCI-3140; N/a –not available.

and full-scale error γ_S. The set of sensors forms a data array, which is necessary, for example, for a control system. In view of the influence on the sensor of a significant number of destabilizing factors, which deform its transformation characteristic and result in occurrence of casual errors, the sensor's transformation error is distributed according to the unbiased normal law at which

$$\delta_{max} \leq \pm 3\sigma_{sensor}, \tag{3.4}$$

where δ_{max} and σ_{sensor} is the limit of sensor error and mean-root square error accordingly.

The second measuring conversion is analog–digital and accompanied by static and dynamic errors. In case of time interval measurements according to the classical indirect counting method (pulse duration, one or several periods of the sensor frequency), the static error is caused by the instability of the reference frequency f_0, the inaccuracy of the time interval separation (the trigger error) and its quantization. In turn, the dynamic error is caused by the measurand changing during the time of measurement as well as one or several periods T_x and measuring cycles, especially in data acquisition systems using time-division channelling and a large number of sensors.

The limit error δ_{max} and the mean-root square error σ_T values of the static error for the period (time interval)-to-code converter is determined by the following way:

$$\delta_{T\,max} = \delta_{Trigger\_error\_max} + \delta_{0\,max} + \delta_{q\,max} \tag{3.5}$$

$$\sigma_T = \sqrt{\sigma_{Trigger\_error}^2 + \sigma_0^2 + \sigma_q^2}, \tag{3.6}$$

where $\delta_{Trigger\_error}$ and $\sigma_{Trigger\_error}$, $\delta_{0\,max}$ and σ_0, $\delta_{q\,max}$ and σ_q are relative, the limit and the mean-root square for trigger error, the reference frequency f_0 and the quantization error accordingly.

Components of the common error of the direct counting method for frequency measurements in many respects are similar to those considered above except for the absence of the trigger error, so

$$\delta_{f\,max} = \delta_{0\,max} + \delta_{q\,max} \tag{3.7}$$

$$\sigma_f = \sqrt{\sigma_0^2 + \sigma_q^2}, \tag{3.8}$$

In multichannel data acquisition systems multiplexers and communication lines between frequency output sensors and the computer as a rule do not influence the sensor's output frequency.

For reduction of the total error, it is necessary to reduce the weight of its components and, first of all, those with dominant values. In the case of frequency–time measurements, it is a quantization error. Its value is directly dependent on the method of the frequency (period)-to-code conversion. This is why the correct choice of a conversion method at the creation of data acquisition systems for frequency–time-domain sensors is one of the main tasks.

The central computer, processing the received data, carries out various mathematical calculations and forms the results array necessary for solution of different control and measuring tasks. This is the so-called third measuring conversion. In view of the above, the mean-root square error of the measuring channel in a data acquisition system σ_{DAQ} can be determined according to the following equation:

$$\sigma_{DAQ} = \sqrt{\sigma_{sensor}^2 + \sigma_{F/C}^2 + \sigma_{Calc}^2}, \tag{3.9}$$

where $\sigma_{F/C} = \sigma_f$ or $\sigma_{F/C} = \sigma_T$ dependent on the method of the frequency-to-code conversion; σ_{calc} is the mean-root square calculating error, carried out by a central computer. In the case of digital output smart sensors with architectures similar to those shown in Figures 3.5–3.7 the sensor error will already include the frequency (period)-to-code conversion error.

3.5   Data Transmission and Error Protection

Data transmission in the digital form from a digital output smart sensor to a remote central computer demands additional measures for error protection of transmitted data. It can be achieved by means of protective coding using various redundant codes. Very frequently cyclic codes are used. According to this, the common coding algorithm can be presented as: $f_{xi}(T_{xi}) \rightarrow N_{parallel} \rightarrow N_{serial} \rightarrow N_{cyclic\,code}$. The cyclic (n, n_{info}) code is the protective code each code combination of which is expressed as a polynom, having a degree not exceeding $(n - 1)$ and divided on the generating multinomial (polynom) $P(x)$ of $n_c = n - n_i$ degree, where n_c is the number of check bits, n_i is the number of information bits and n is the code block length. The checking by the cyclic redundant code represents a method similar to the method of the control sum calculation, differing in that the cyclic redundant code has, as a rule, a two-byte length and is calculated based on the polynom of the divider. The cyclic code combination represents a 16-bit remainder of the division. Using such a method, at each moment of time, a block of a certain size is transmitted. Any transferred bit influences the cyclic code, therefore such an error check is most effective among similar methods.

The novel method of noise-resistant signal transmission in which the number of detected or corrected errors are increased twice is described below [91].

The complex solution of this task becomes considerably complicated for increasing transmission rate and length of code combinations. It especially appears by the creation of precision multichannel data acquisition systems, remote control and metering systems and digital wireless sensors and transducers. Such systems require increased correcting ability of the code without increase of the redundancy [92]. It requires the development of non-traditional methods, coders and decoders in view of factors influential to the choice of the correcting code, signal modulation, transmission rate and simplicity of realization.

3.5.1   Essence of quasi-ternary coding

The data coding depends on the right choice of method, which should ensure smaller redundancy of the correcting code at equal noise stability. Unlike conventional coding, quasi-ternary coding doubles the number of detected or corrected errors without a change in the number of check bits n_c of the binary correcting code. The increase in the correcting ability of the quasi-ternary code is achieved by using signal redundancy, in addition to code redundancy. The code conversion is carried out with the help of the phase manipulator forming two various informative tags at '1' and '0' transmission of the multidigit codeword by digit-by-digit phase manipulation. This favourably differs from the absolute phase and relative phase modulation, offered by Professor N. T. Petrovich. According to the algorithm of digit-by-digit phase manipulation, the difference of phases of the carrier wave of the code signal (meander or harmonic) between any adjacent code elements is always 180^0. Thus all bits of the code are transmitted: '1' by the phase manipulations of one polarity, for example $+180^0$, and '0' by the opposite polarity [88]. It is possible also to use other types of manipulation, for example, the polar meander carrier wave or frequency, however, they concede to digit-by-digit phase manipulation.

The quasi-ternary serial code with the information n_i and check bits n_c of the correcting code that is transmitted by the digit-by-digit phase-manipulation signal is formed as a result of the code conversion. The code is self-clocked. It has the minimum value of the constant component and twin elements with the single or binary phase manipulation carrier wave for the denotation of the start and end of the codeword. The duration of its bits elements is a variable in bounds from 2.5 up to 3 periods T of carrier wave depending on the value of adjacent bits of the code [91−93].

Concerning correcting codes, it is expedient to use linear block cyclic separable codes. Unlike the other systematic codes, they correct not only single (independent), but also the group errors. Such codes are simpler in implementation, easily allow the introduction of redundancy at coding, if the level is insufficient, and select an informative part of the block at their decoding. This class of codes contains such codes as Bose−Chaudhuri−Hocquenghem (BCH) code and Golay code [94]. The algebraic theory for construction of block codes as well as coding and decoding algorithms were developed for them. The coding efficiency is determined by their redundancy, which can be varied in a wide range. The quasi-ternary coding can also be used for Hamming codes, group codes, iterated codes, codes with repetition and inversion, etc.

3.5.2   Coding algorithm and examples

Traditionally, the following operations are executed for coding by the separable cyclic code with a generating polynomial $P(x)$ of x^{nc} degree of the information group from the n_i bits of the polynomial $G(x)$: the symbol multiplying $G(x)$ on x^{nc}; the division of the obtained product on the polynomial $P(x)$; the summation of the obtained residual $R(x)$ with the product x^{nc}. The obtained polynomial

$$F(x) = x^{nc}G(x) \oplus R(x) \qquad (3.10)$$

contains the n bits. It has a necessary redundancy determined by its check bits $n_c = n - n_i$, and represents the polynomial of the separable systematic cyclic code, traditionally formed at transmission of the polynomial $G(x)$ with the help of linear code filters with the shift registers and feedbacks. Thus, the following identity is valid:

$$F(x) \equiv 0 \ \mathrm{mod} P(X), \qquad (3.11)$$

indicating on the long division without the residual of the polynomial $F(x)$ on the generating polynomial $P(x)$. The identical is also used for decoding. All these operations are absent in the offered algorithm. In this case, the code filters are not used because the coder is performed by the ROM, which is a generator of check bits of the residuals $R(x)$ of the unit matrixes I_{ni}. The remainder $R(x)$ required for the polynomial $G(x)$ is formed only as a result of the summation to modulo 2. By the hardware realization the clock pulses counter controls the coder. The capacity of the counter is determined by the number of information bits n_i. The values of check bits of the residual $R(x)$ are determined by the n_c adders. The ROM is used for the storage and bit-by-bit address reading of the residuals of appropriate rows in the unit matrix I_{ni} of the polynomial $G(x)$. Depending upon the information bit ('0' or '1') in the decoder's input, the reading of the residual of a row in the unit matrix I_{ni} of the given bit is forbidden or permitted. The summarization of the residuals at transmission proceeds until all information bits have arrived. Then the residual $R(x)$, appropriate to one from 2^{ni} values of the transmitted binary code is generated. Examples of coding that is carried out with the help of the above algorithm are shown below.

Example 1. Coding of the polynomial $G(x)$ by the Golay code with parameters (23, 12) and the generating polynomial $P(x) = x^{11} + x^9 + x^7 + x^5 + x + 1 = 101011100011$, at $n = 23$ bits; $n_i = 12$ bits; $n_c = 11$ bits; $d = 7$ is the code distance; $s = 3$ is the number of correctable errors; $r = 6$ is the detectable error).

The generating matrix $C(23, 12)$ is the following:

$$
C(23, 12) = \left|
\begin{array}{cccccccccccc|ccccccccccc}
0 & 0 & 0 & 0 & 0 & 0 & 0 & 0 & 0 & 0 & 0 & 1 & 0 & 1 & 0 & 1 & 1 & 1 & 0 & 0 & 0 & 1 & 1 \\
0 & 0 & 0 & 0 & 0 & 0 & 0 & 0 & 0 & 0 & 1 & 0 & 1 & 0 & 1 & 1 & 1 & 0 & 0 & 0 & 1 & 1 & 0 \\
0 & 0 & 0 & 0 & 0 & 0 & 0 & 0 & 0 & 1 & 0 & 0 & 0 & 0 & 1 & 0 & 1 & 1 & 0 & 1 & 1 & 1 & 1 \\
0 & 0 & 0 & 0 & 0 & 0 & 0 & 0 & 1 & 0 & 0 & 0 & 0 & 1 & 0 & 1 & 1 & 0 & 1 & 1 & 1 & 1 & 0 \\
0 & 0 & 0 & 0 & 0 & 0 & 0 & 1 & 0 & 0 & 0 & 0 & 1 & 0 & 1 & 1 & 0 & 1 & 1 & 1 & 1 & 0 & 0 \\
0 & 0 & 0 & 0 & 0 & 0 & 1 & 0 & 0 & 0 & 0 & 0 & 0 & 0 & 1 & 1 & 0 & 0 & 1 & 1 & 0 & 1 & 1 \\
0 & 0 & 0 & 0 & 0 & 1 & 0 & 0 & 0 & 0 & 0 & 0 & 0 & 1 & 1 & 0 & 0 & 1 & 1 & 0 & 1 & 1 & 0 \\
0 & 0 & 0 & 0 & 1 & 0 & 0 & 0 & 0 & 0 & 0 & 0 & 1 & 1 & 0 & 0 & 1 & 1 & 0 & 1 & 1 & 0 & 0 \\
0 & 0 & 0 & 1 & 0 & 0 & 0 & 0 & 0 & 0 & 0 & 0 & 1 & 1 & 0 & 0 & 0 & 1 & 1 & 1 & 0 & 1 & 1 \\
0 & 0 & 1 & 0 & 0 & 0 & 0 & 0 & 0 & 0 & 0 & 0 & 1 & 1 & 0 & 1 & 0 & 0 & 1 & 0 & 1 & 0 & 1 \\
0 & 1 & 0 & 0 & 0 & 0 & 0 & 0 & 0 & 0 & 0 & 0 & 1 & 1 & 1 & 1 & 1 & 0 & 0 & 1 & 0 & 0 & 1 \\
1 & 0 & 0 & 0 & 0 & 0 & 0 & 0 & 0 & 0 & 0 & 0 & 1 & 0 & 1 & 0 & 1 & 1 & 1 & 0 & 0 & 0 & 1 \\
\end{array}
\right|
$$

$$\underbrace{\qquad\qquad}_{I_{ni}^T = I_{12}^T} \qquad \underbrace{\qquad\qquad}_{C_{ni,nc} = C_{12,11}}$$

$$(3.12)$$

The coding of the polynomial $G(x) = 111000011001$ at transmission by the quasi-ternary code is shown below.

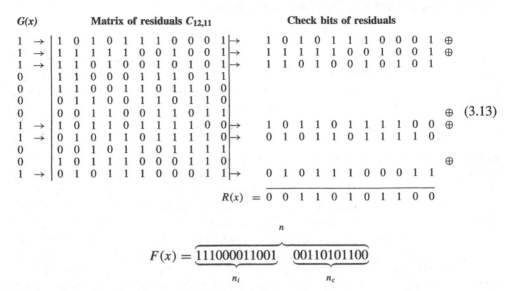

$$F(x) = \overbrace{\underbrace{111000011001}_{n_i} \quad \underbrace{00110101100}_{n_c}}^{n}$$

It is easily proved that the identity (3.11) will be realized by division of the generated polynomial $F(x)$ by the generating polynomial $P(x)$.

The time diagram for the generated digit-by-digit phase-manipulated signal of the noise-resistant quasi-ternary code at the start-stop transmission mode using a meander or harmonic carrier wave is shown in Figure 3.8. The Golay code with parameters (23, 12), applied in the example is one of a few nontrivial perfect binary codes [94].

There is also the so-called extended Golay code with parameters (24, 12) and with higher efficiency ($d = 8$; $s = 3$; $r = 7$). The coding is realized similarly to Equation (3.13). It is necessary to provide an additional check bit, determined by the polynomial $P(x) = x^9 + x^8 + x^5 + x^4 + x^2 + x$.

Example 2. Coding of the polynomial $G(x) = 1110\,001$ by BCH code with parameters (15,7), that is given by the generating polynomial $P(x) = x^8 + x^7 + x^6 + x^4 + 1$.

After construction of the generating matrix $C_{n,ni}$ for the BCH code (15,7) $C_{15,7} = |I_7 T||C_{7,8}|$ using the residual matrix $C_{7,8}$, let us form the code combination $F(x)$ of the BCH code for the polynomial $G(x) = 1110\,001$ transmission companion to Equation (3.13):

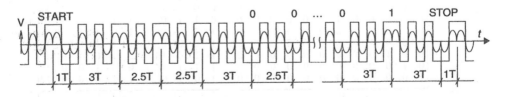

Figure 3.8   Digit-by-digit phase-manipulated quasi-ternary coding

$G(x)$		$C_{7,8}$								Check bits of residuals								
1	→	1	1	1	0	1	0	0	0	→	1	1	1	0	1	0	0	0	⊕
1	→	0	1	1	1	0	1	0	0	→	0	1	1	1	0	1	0	0	⊕
1	→	0	0	1	1	1	0	1	0	→	0	0	1	1	1	0	1	0	
0		0	0	0	1	1	1	0	1										
0		1	1	1	0	0	1	1	0										
0		0	1	1	1	0	0	1	1									⊕	
1	→	1	1	0	1	0	0	0	1	→	1	1	0	1	0	0	0	1	

$$(3.14)$$

$$R(x) = 0\ 1\ 1\ 1\ 0\ 1\ 1\ 1$$

$$F(x) = \underbrace{1110001}_{n_i},\ \underbrace{01110111}_{n_c}$$

It is easy to prove that the identity (3.11) is true. The generated code $F(x)$ has the code distance $d = 5$ ($s = 2$ or $r = 4$). The quasi-ternary code, obtained from the BCH code (15,7) after decoding will ensure $s = 4$ in the error-checking mode and $r = 8$ in the error-detection mode. The time diagram of the generated quasi-ternary code for the first five bits of the polynomial $G(x)$ coincides with that mentioned above. The signal redundancy increases the coding efficiency without increasing BCH code redundancy.

3.5.3 Quasi-ternary code decoding

The reception of the self-clocking digit-by-digit phase-manipulated signal of n-bit quasi-ternary correcting code is carried out by two non-linear phase selectors. The selectors are quasi-optimal integrated receivers, which differ by a small time response. The probability of the error is determined as

$$P = 0.5[1 - F(N\sqrt{2})].$$
$$(3.15)$$

In the case of the harmonic carrier wave we have

$$N = \frac{V_{\text{eff\_signal}}}{V_{\text{eff\_noise}}}\sqrt{\frac{\Delta F}{B}}$$
$$(3.16)$$

where: $V_{\text{eff\_signal}}$ and $V_{\text{eff\_noise}}$ are the effective voltage of the signal and noise accordingly; ΔF is the transmission bandwidth of the input filter; B is the signal manipulation speed.

Two code combinations are formed from the receipted code combination of the quasi-ternary cyclic code in the inputs of phase selectors. The first combination $F_1'(x)$ contains only impulses of positive phase manipulation '1', and the second $F_0'(x)$, only impulses of the negative phase manipulation '0'. Each of them can be true, i.e. to coincide with the transmitted code combination as well as false because of the noise effect. The decoding according to the algorithm, assigned by the decoder, will be realized simultaneously with the bit-by-bit input and storage of the n-bits polynomials $F_1'(x)$ and $F_0'(x)$.

In the case of decoding with error correction the division will be realized with the aim of determining the identity (3.11). If some errors are present in one or both code combinations, the following identities with allowance (3.11) will be used:

$$F_1'(x) \bmod P(x) \equiv E_{1i}'(x) \bmod P(x) = \delta_{1i}(x), S_{1i}(x) \tag{3.17}$$

$$F_0'(x) \bmod P(x) \equiv E_{0i}'(x) \bmod P(x) = \delta_{0i}(x), S_{oi}(x) \tag{3.18}$$

where $\delta_{1i}(x)$ and $\delta_{0i}(x)$ are the syndromes of error vectors $E_{1i}(x)$ and $E_{0i}(x)$, arising in the polynomials $F_1'(x)$ and $F_0'(x)$, that do not depend on the transferred polynomial $F(x)$ and are determined only by the position of error in the code combination. All methods for error-position detection with the consequent correction are based on the analysis of syndromes $S_{1i}(x)$ and $S_{0i}(x)$. Its realization becomes considerably according to increasing number of corrected errors. Therefore the number S is restricted by the inequality $S \leq 3$.

Decoding by the error detection mode is simpler, as it is based on the sum to modulo 2 and comparison. As more expedient, only this mode is further considered, especially in the case of codes with high degree polynomials $P(x)$ and large code distance $d_{\min}$. It is stipulated by the following factors:

- The detecting ability of the cyclic code at the same redundancy is higher than the correcting ability ($r = 2s$).

- The number of logical and other operations that should be executed by the decoder is significantly less in comparison with the number of operations for their correction.

- The reliability of decoding is higher, as for the given code efficiency it is possible to reduce the code distance $d_{\min}$ and with allowance the quasi-ternary coding, the check bits decrease twice.

For this reason decoding in the detection mode is widely used in different systems:

- With the make-decision feedback, where the BCH code with generating polynomial $P(x) = x^{16} + x^{12} + x^5 + 1$ is used.

- Without the feedback, but using the reception with erasure and the possibility of restoring the erased code units, due to its redundancy.

The analysis of the noise stability has shown that decoding in the error-detection mode is more preferable for independent errors with the probability $p = 10^{-5} - 10^{-3}$ as the noise stability increases up to $2 - 3$ orders for the cyclic with $d = 3$ and up to $3 - 4$ orders for the code with $d = 4$.

The decoding algorithm contains operations, which are executed up to the termination of arrival of the phase-manipulated signal of the quasi-ternary cyclic code in the following sequence:

1. Creation from the received n-bits code combination of the polynomial $F'(x)$ the polynomials $F_1'(x)$ and $F_0'(x)$ containing only '1' or '0' accordingly, recording in the buffer register the $F_1'(x)$ and $F_0'(x)$ after inverting bits.

2. Selection from $F_1'(x)$ and $F_0'(x)$ the informative n_i bits polynomials $G_1'(x)$ and $G_0'(x)$ and creating according to the algorithm (3.13) the check bits of appropriate strings of the unit matrix I_{ni} of the residuals $R_1'(x)$ and $R_0'(x)$, stored in the ROM.

3. Selection of the received polynomials of residuals and their comparison with the residuals $R_1'(x)$ and $R_0'(x)$ by execution of the following operations:

- the equivalence and signal forming for the reading of the polynomial $G(x)$ from the buffer register at the moment of arrival of the 'STOP' signal (Figure 3.8)
- the non-equivalence and signal forming for prohibiting the reading of the polynomial $G(x)$ because of errors in the received code combination.

The full coincidence of compared polynomials of residuals is informed by the signal of the absence of error ($ER = 0$).

This novel method of the efficiency rise for coding without increasing the code redundancy of the used code by use of signal redundancy has confirmed the high efficiency in practice. The additional increase of the code redundancy considerably increases the detecting ability of the cyclic code. Searches for the algorithm simplification for the error correction in code combinations of the quasi-ternary code (Patents No.107 54 37, 1146 788, 1124 363 (USSR), [88]) are continuing.

The use of PGA or FPGAs allows multifunctional programmed devices containing the check bits of matrixes $I_{ni,nc}$. Thus, the choice of the required code redundancy for many perspective codes can be realized. The offered non-traditional method of coding promotes the further rise of the coding efficiency by using generating polynomials $P(x)$ of high degrees−14, 16, 24 and 32.

Summary

The method of data acquisition with time-division channelling allows inexpensive multichannel data acquisition systems for quasi-digital sensors.

The method of data acquisition for quasi-digital sensors with space-division channelling is becoming increasingly attractive due to the low cost of frequency-to-code converters. A separate frequency-to-code converter for each channel realizes a much higher polling frequency for each channel.

The accuracy of modern industrial DAQ boards for frequency−time parameters of electric signals often does not allow using them together with modern precision frequency (period, duty-cycle, time interval) output sensors. Also, it is desirable to have an opportunity to change both the accuracy and the measurement time directly during data acquisition.

The quantization error has the essential influence on the data acquisition system's accuracy. Its value is directly dependent on the method of the frequency (period)-to-code conversion. That is why the correct choice of a conversion method by creation of data acquisition systems for frequency−time-domain sensors is one of the main tasks of the system design.

Data transmission in the digital form from a digital output smart sensor to a remote central computer demands additional measures for the error protection of transmitted data. It can be achieved by means of protective coding using of various redundant codes. Very frequently cyclic codes are used for this aim.

Taking into account individual features of the chosen correction methods, transmission devices and specificity of system application, developers of data acquisition system should solve which requirements the coding algorithm would best meet.

4

METHODS OF
FREQUENCY-TO-CODE
CONVERSION

One of the main parts of frequency–time-domain smart sensors is the frequency (period, duty-cycle or time interval)-to-code converter. This unit directly influences such sensor metrological characteristics, as the accuracy and the conversion time as well as the power consumption. In spite of the fact that the frequency can be converted into digital code more precisely in comparison with other informative parameters of the signal, in practice, it is not a trivial task of simple time-window counting. Besides, very often a consumer or a sensor manufacturer is not an expert in the area of frequency-time measurements. Let us consider the most popular frequency-to-code conversion discrete methods and give an analysis of metrological performances, conversion frequency ranges and requirements for realization with the aim of choosing the most-appropriate conversion method.

The formation and development of this perspective began after the publication in 1947 of the first patent on the electronic ADC with narrow time intervals offered by Filipov and Negnevitskiy in 1941. This converter was based on the discrete counting method of normalized impulses of the reference frequency (Patent 68785, USSR).

Today, there are many patents concerning various conversions and measuring methods of frequency–time parameters of electrical signals. The main methods are:

- The standard counting method for measurement of the average frequency for a fixed reference gate [95–97]. Due to the advantages and simplicity of implementation the method is still used today [98].

- The indirect method of measurement of the instantaneous frequency for the time-interval measurement [99,100].

- The interpolation method (with analog and digital interpolation) [97,101,102].

- The method of recirculation [97].

- The method of the single or multiple vernier. Many new methods for frequency measurement, the phase and other values as well as their ratios were developed on the basis of this method [103].

- The method of delayed coincidences and additional channels [100,103].

- The reciprocal counting method or the so-called coincidence measuring method [97–98,104,105] (this method was used in the Hewlett-Packard counter HP5345A) and its modifications: double casual coincidence of impulses and impulse packets.

- Special methods based on the first two methods ensuring: direct measurements of values, including the measurement of low and infralow frequencies [95,96,106]; the tracing mode [95,103]; the weight average [107]; the measurement of frequency, commensurable with the interference [96,107]; the increase of static and dynamic accuracy [4,108,109]; the measurement of instantaneous frequencies by the signal derivation [96]; the frequency range extension, the parametric adaptation and other capabilities.

- The high resolution and high accuracy M/T method [110].

- The method with constant elapsed time (CET method) [111].

- The double buffered measurement method [112].

- The DMA transfer method [113].

- The accurate method of the frequency-time measurement with the non-redundant reference frequency [114].

- The method of the dependent count (MCD) [115,116].

Naturally, for different reasons, not all the listed methods may be used for frequency-to-code conversion in smart sensors. In this chapter, we shall consider classical methods of frequency (period)-to-code conversion.

4.1   Standard Direct Counting Method (Frequency Measurement)

The frequency counting scheme shown in Figure 4.1 is one of the most commonly used techniques for converting the output of a sensor to a numerical quantity. The rest of the circuitry, which must be used to reset the counter before the next gate period occurs, is not shown in this simplified diagram.

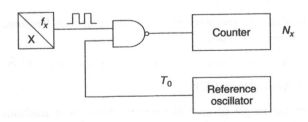

Figure 4.1   Simplified diagram of standard counting method

The conversion method consists of counting a number of periods T_x of the unknown frequency f_x during the gate time (the reference interval of time) T_0. The gate time is formed by dividing a reference frequency f_0 down to a suitable period. The output pulses of the sensor are simply accumulated during the time the gate signal is high. The result of conversion can be calculated by the following way:

$$N_x = T_0/T_x = T_0 f_x \qquad (4.1)$$

If T_0 is equal to one second, for instance, the output N_x is equal to the sensor frequency f_x. In the common case, the converted frequency is determined according to the following equation:

$$f_x = N_x \cdot f_0 = \frac{N_x}{T_0} \qquad (4.2)$$

The time diagram of the described circuit (Figure 4.2) shows that the absence of synchronization of the beginning and end of the gate time T_0 with pulses f_x results in an error of measurement; its absolute value is determined by values Δt_1 and Δt_2. It is easy to see, that actually the measurement time is

$$T_0' = N_x \cdot T_x = N_x/f_x = T_0 + \Delta t_1 - \Delta t_2 \qquad (4.3)$$

therefore

$$N_x = T_0 \cdot f_x + (\Delta t_1 - \Delta t_2)/T_x \qquad (4.4)$$

$$T_0 = N_x T_x - \Delta t_1 + \Delta t_2 = N_x T_x \pm \Delta t = N_x T_x \pm \Delta_q \qquad (4.5)$$

Time intervals Δt_1 and Δt_2 can be changed independently of each other, accepting values from 0 up to T_x with equal probability. Then the maximum relative quantization (discretization) error caused by the absence of synchronization, will be

$$\delta_q = \pm \frac{1}{N_x} = \pm \frac{1}{T_0 \cdot f_x} \qquad (4.6)$$

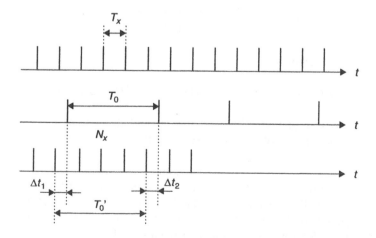

Figure 4.2   Time diagrams of direct counting method

Without special measures the absolute error Δ_q, with a maximal value that does not exceed ± 1 count pulse, will be distributed according to the triangular law (Simpson's distribution law)

$$W(\Delta_q) = \begin{cases} 0, & at - 1 > \Delta_q > 1 \\ 1 + \Delta_q, & at - 1 \le \Delta_q \le 0 \\ 1 - \Delta_q, & at \ 0 \le \Delta_q \le 1 \end{cases} \tag{4.7}$$

with the mathematical expectation (mean), the dispersion and the mean-square deviation:

$$M(\Delta_q) = 0; \ D = 1/6; \ \sigma(\Delta_q) = \pm\sqrt{D} = \pm 1/\sqrt{6} \tag{4.8}$$

accordingly.

The mean-root square error $\sigma(\Delta_q)$, because of the distribution law symmetry $W(\Delta_q)$ corresponds to the standard deviation.

The moment of the beginning of measurement can be synchronized with the converted frequency but the moment of the ending of measurement cannot be synchronized. The quantization error for this case may be determined from Equation (4.4), if $\Delta t_1 = 0$. Then the quantization error will be determined according to the following equation:

$$\delta_q = -\frac{1}{T_0 \cdot f_x} \tag{4.9}$$

The value δ_q can be reduced, if f_x pulses are shifted compulsorily in a half-period in reference to the beginning of the gate time T_0. In this case $\Delta t_1 = T_x/2$, and the quantization error can be determined

$$\delta_q = \pm\frac{1}{2T_0 \cdot f_x} \tag{4.10}$$

In this case, the distribution of the methodical quantization error is the uniform symmetric (unbiased) law

$$W(\Delta_q) = \begin{cases} 0, & at - 0.5 > \Delta_q > 0.5 \\ 1, & at - 0.5 \le \Delta_q \le 0.5 \end{cases} \tag{4.11}$$

with numeric characteristics

$$M(\Delta_q) = 0; \ D = 1/12; \ \sigma(\Delta_q) = \pm\sqrt{D} = \pm 1/2\sqrt{3} \tag{4.12}$$

Distribution laws (Equations (4.7) and (4.11)) and their numerical characteristics are valid, if the duration τ of pulses f_x satisfies the following condition $\tau \ll T_{xmin}$. Otherwise, Simpson's law is transformed into the biased trapezoidal distribution law, the uniform unbiased law into the biased law, and the value of error Δ_q is increased.

It is obvious, that by the calculation of the gate time T_0 according to Equation (4.5) its maximum value will be determined by the lower frequency $f_{x\,min}$ of the conversion frequency range. The high range of frequency conversion is limited by the maximum pulse counter speed. So, for example, the microcontroller Intel D87C51AF has a 3 MHz counting frequency.

By the direct frequency measurement, there are two essential error components. These are the frequency reference error δ_{ref} and the above considered, quantization error δ_q.

The frequency reference error δ_{ref} is the systematic error, caused by inaccuracy of the initial tuning and the long-term instability of the quartz generator frequency. The casual component of the resulting error is determined by the short-term instability of the frequency f_0.

As known, the frequency deviation of the non-temperature-compensated crystal oscillator from the nominal due to the temperature change is $(1-50) \times 10^{-6}$ in the temperature interval $-55-+125\,°C$. This error's component may be essential in the measurement of high frequencies. For reduction of the systematic frequency reference error an oven-controlled crystal oscillator is used at which the maximal value $\delta_{ref} = 10^{-6}-10^{-8}$ remains in the given limits for a long time.

The maximum permissible absolute quantization error is

$$\Delta_q = \pm f_0 = \pm \frac{1}{T_0} \tag{4.13}$$

In turn, the limits of the absolute error of frequency-to-code converters based on the standard direct counting method can be calculated as:

$$\Delta_{max} = \pm \left(\delta_{ref} f_x + \frac{1}{T_0} \right) \tag{4.14}$$

Limits of the relative error of such a frequency-to-code converter in percent is

$$\delta_{max} = \pm \left(\delta_{ref} + \frac{1}{f_x \cdot T_0} \right) \cdot 100 \tag{4.15}$$

The quantization error δ_q depends on the converted frequencies. It is negligible for high frequencies above 10 MHz, grows at the frequency reduction and may reach an inadmissible value in the low and infralow frequency range. For example, for $f_x = 10$ Hz at gate time $T_0 = 1$ s the quantization error will be 10%. On the other hand in order to reach a reasonable quantization error of at least 0.01%, a frequency of 10 Hz will result in an increased conversion time up to $T_0 = 1000$ s. Effective methods for reducing quantization error in the standard direct counting method are:

(a)   multiplication of converted frequencies f_x in k times and subsequent measurement of the frequency $f_x \cdot k$, that is reduced to increase the pulse number inside the gate time

(b)   using weight functions.

Both ways result in some increase in hardware or increase in the chip area for the realization of the additional frequency multiplier, increase of the conversion time by the realization of the weight averaging with the help of the microcontroller core.

One of the demerits of this classical method is the redundant conversion time in all frequency ranges, except the nominal frequency.

4.2   Indirect Counting Method (Period Measurement)

The indirect counting method is another classical method for frequency-to-code conversion. This method is rather effective for the conversion of low and infralow frequencies. According to this method the number of pulses of the high reference frequency f_0 is counted during one T_x or several n periods T_x (Figure 4.3).

Thus, the following number will be accumulated in the counter:

$$N_x = n \cdot T_x/T_0 = n \cdot f_0/f_x, \tag{4.16}$$

where n is the number of periods T_x. The number N_x is equal to the converted period. By the same way, it is possible to convert the pulse width t_p, or the time interval τ between start and stop pulses into the code.

The number of pulses N_x counted by the counter is determined by the number of periods $T_0 = 1/f_0$ during the time interval T_x. Therefore the average value of T_x is equal to:

$$T_x = N_x \cdot T_0 \tag{4.17}$$

In view of the absolute quantization error because of non-synchronization

$$T_x = (N_x - 1)T_0 + \Delta t_1 + (T_0 - \Delta t_2) = N_x T_0 + \Delta t_1 - \Delta t_2 = N_x T_0 \pm \Delta_q \tag{4.18}$$

Therefore, distribution laws of these errors $W(\Delta_{t1})$ and $W(\Delta_{t2})$ are equiprobable and asymmetrical with the probability $1/T_0$ and with the mean

$$M(\Delta t_1) = 0.5 \cdot T_0, \, M(\Delta t_2) = -0.5 \cdot T_0 \tag{4.19}$$

The random additive methodical quantization error Δ_q is determined by the sum of the independent and distributed errors according to the uniform distribution law random errors Δt_1 and Δt_2. The maximum value of the error is $\Delta_{xq\,max} = \pm T_0$. It is

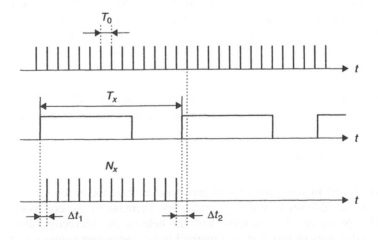

Figure 4.3   Time diagrams of indirect counting method

distributed according to the triangular (Simpson's) distribution law $W(\Delta_q)$ with the null mathematical expectation:

$$M(\Delta_q) = M(\Delta t_1) + M(\Delta t_2) = 0 \tag{4.20}$$

and the dispersion

$$D(\Delta t_1) = D(\Delta t_2) = D(\Delta t) = \frac{T_0^2}{12} \tag{4.21}$$

Because of the distribution law symmetry $W(\Delta_q)$ the mean-root square error coincides with the standard deviation

$$\sigma(\Delta_q) = \sqrt{\sigma^2(\Delta t_1) + \sigma^2(\Delta t_2)} = \sqrt{\frac{T_0^2}{12} + \frac{T_0^2}{12}} = \pm\frac{T_0}{\sqrt{6}} \tag{4.22}$$

By synchronization of the reference frequency f_0 with the beginning of the converted interval T_x, it is possible to provide $\Delta t_1 = 0$. However, it is impossible to synchronize the ending of this interval in a similar way. Therefore, the methodical error Δ_q remains irremovable in principle. Owing to such synchronization the symmetric uniform distribution law $W(\Delta_q)$ is transformed into the asymmetrical uniform law with the probability $1/T_0$ and the error be further increased. Therefore

$$\Delta_{q\,max} = \Delta t_2 = T_0; \ M(\Delta_q) = -0.5T_0; \ \sigma(\Delta_q) = \frac{T_0}{\sqrt{3}} \tag{4.23}$$

Because of the occurrence of the systematic error component $-0.5T_0$ the standard deviations will be less than twice the mean-root square errors $\sigma(\Delta_q)$. Additionally shifting the reference pulses to $0.5T_0$, we receive the uniform symmetric distribution law.

$$W(\Delta_q) = \begin{cases} 0 & at -0.5T_0 > \Delta_q > 0.5T_0 \\ \dfrac{1}{T} & at -0.5T_0 \leq \Delta_q \leq 0.5T_0 \end{cases} \tag{4.24}$$

with the numerical characteristic

$$M(\Delta_q) = 0; \ D = \frac{T_0}{12}; \ \sigma(\Delta_q) = \pm\sqrt{D} = \pm\frac{T_0}{2\sqrt{3}} \tag{4.25}$$

and the standard deviation coincides with the values of $\sigma(\Delta_q)$. We notice that this error is distributed according to the uniform law each time the measurand, with a continuous value, is replaced by a discrete value.

In order that the result equals the converted frequency it is necessary to calculate the ratio

$$N_{fx} = \frac{1}{N_x} \tag{4.26}$$

By using the microcontroller core, this operation is carried out without any problems and in parallel to the conversion process. It is also possible to use various functional converters or digital integrators working in the mode of hyperbolic function modelling so that the conversion results in units of frequency. Among a significant number of specially developed methods of instant frequency conversion and the reproduction of the inversely proportional dependence the *single-cycle* and *bicyclic* methods were favourably allocated [95,96].

We shall specify below only the most distinctive features and technical performances of the best-developed devices, defined by their metrological characteristics and applied to frequency-to-code converters.

Converters on the basis of *single-cycle* methods have the greatest speed and accuracy. The linearization in them is realized by:

- an integrator with parallel function, carrying out the pulse modelling of the hyperbolic dependence simultaneously with the quantization of the period T_x (or the multiple time interval to its duration)

- a decoder based on a read-only memory (ROM), converting the code of the number N_{Tx} into the code of the result N_{fx}.

In the first case, the approximation of dependence is realized with high accuracy without added complexity, in comparison with piecewise-linear and piecewise-exponential approximations. As the number in the decrement counter of the integrator is changed continuously, then the quantization frequency of the period Tx will also be changed continuously. It provides the $\varphi(T_x)$ reproduction with the best accuracy. In measurements of the frequency of rectangular pulses, the methodical linearization (approximations) error is negligibly small. The usage of the integrator, by using a small frequency changes, incorporates a tracking measuring mode instead of a cyclic mode.

In the second case at the end of the quantization period T_x the code N_{fx} is formed by the reading pulse on the received code of the number N_{Tx} in the decoder. The methodical linearization error is absent, because the table of conformity $N_{Tx} \rightarrow N_{fx}$ is made on the basis of the formula $N_{fx} = N_S/N_{Tx}$, and each number N_{fx} is rounded.

Converters using *bicyclic* methods differ in a shorter speed. In the first step one or n periods are quantized. At the end of this step the number of pulses N_{Tx} is formed in the counter. During the second step, the $N_{Tx} \rightarrow N_{fx}$ conversion is realized by repeated addition or subtraction:

- the number N_{Tx} for calculation $N_S = N_{Tx} \cdot N_{fx} = const$ (addition) or $N_{fx} = N_S/N_{Tx}$ (subtraction), where Ns is the constant maximum number determined by the given accuracy and the time of $N_{Tx} \rightarrow N_{fx}$ conversion

- two pulse sequences: one uniform with the constant clock frequency controlled by the dynamic range $D_f = f_{x\,max}/f_{x\,min}$, and the other non-uniform from the digital feedback loop with average frequency determined by the complement code of the binary number N_{Tx}.

In the first case, multiplying-dividing operations are carried out. Therefore, for simplification of the circuit realization (by the hardware realization) integrators with sequential carry (the so-called binary frequency multipliers) are used. Thus the multiplication is carried out by the consecutive addition of the number N_{Tx} with itself up to a given constant number Ns, and division by the consecutive subtraction from N_s the numbers N_{Tx} up to the required. The number of iterations N_{fx} carried out in both cases determines the result of the measurement of the instant frequency for one or several periods T_x. These methods may also be used for the creation of phase shifts, the off-duty factor, the duty-cycle, etc. to code converters.

In the second case when digital feedback (positive or negative) is used, the consecutive integrator with extended functionality for pulse addition/subtraction to/from the constant clock pulse sequence is also applied. A frequency divider with digital feedback was constructed in order to combine the period quantization and the functional transformation in time. This also reduced the hardware requirements, because the counter was simultaneously working as the frequency divider. Thus, the methodical linearization error and the rounding error of the result occur.

Using the described methods of measurement demands additional digital signal processing that in some cases results in increased hardware. However, the changeover from average frequency measurements to instant frequency measurements was reasonable. It provided increased speed and dynamic error reduction for measurement of fast-changing quantities, for example, in vibration, flow and pressure measurements.

Commonly, the error of frequency-to-code converters of the periodic signal of any form based on the conventional indirect counting method is determined by the instability of the reference frequency, the quantization error and the trigger error due to internal and input signal noises. Trigger errors occur when a time interval of the measurement starts or stops too early or too late because of noise on the input signal. There are two sources of this noise: the noise on the signal being measured and the noise added to this signal by the counter's input circuitry.

The relative quantization error can be calculated according to the following equation:

$$\delta_q = \frac{f_x}{n \cdot f_0} \cdot 100 \tag{4.27}$$

The quantization error can be reduced by increasing the reference frequency f_0 or $n-$numbers of converted periods T_x. It increases with increasing number of converted frequencies. Limits of the relative error of the frequency-to-code converter based on the indirect counting method are

$$\delta_{\max} = \pm \left(\delta_{\text{ref}} + \frac{1}{f_0 T_x n} + \frac{\delta_{\text{TriggerError}}}{n} \right) \tag{4.28}$$

In turn, the trigger error can be calculated as

$$\delta_{\text{TriggerError}} = \frac{1.4\sqrt{(V_{\text{noise−input}})^2 + (V_{\text{noise−signal}})^2}}{S \cdot T_x}, \tag{4.29}$$

where S is the signal slew rate (V/s) at trigger point. At rectangular pulses with the wavefront duration no more than $0.5T_0$ the trigger error is equal to zero. Such output signals are used in the majority of modern frequency output sensors.

Using measurements of time intervals or pulse duration the total error is determined according to the equation

$$\delta_{\max} = \pm \left(\delta_{\text{ref}} + \frac{1}{f_0 T_x} + 2 \cdot \delta_{\text{TriggerError}} + \delta_{\text{TriggerLevelTimingError}} \right) \tag{4.30}$$

In this case, the trigger level error is twice increased because of the start and stop pulse. The trigger level timing error results from the trigger level setting error due to the deviation of the actual trigger level from the set (indicated) trigger level and the amplifier hysteresis if the input signal has unequal slew rates. It can be calculated by the following way:

$$\delta_{\text{TriggerLevelTimingError}} = \frac{\Delta V_{\text{Level1}}}{S_1 \cdot T_x} + \frac{\Delta V_{\text{Level2}}}{S_2 \cdot T_x} \tag{4.31}$$

Usually the inaccuracy of setting levels on the wavefront and the tail are equally accepted ($\Delta V = 20–30$ mV). If their slew is identical it is possible to use the following equation [105]:

$$\delta_{\text{TriggerLevelTimingError}} = \frac{2\Delta V_{\text{Level}}}{S \cdot T_x} \tag{4.32}$$

The quantization error of the frequency-to-code converter using the indirect counting method also depends on the measurand frequency f_x (Equation 4.27).

From the given equations it follows that the conversion of short time intervals causes a large quantization error. It can be reduced by three possible ways. Two of them are obvious direct solutions–increasing of the reference frequency f_0 and converting a

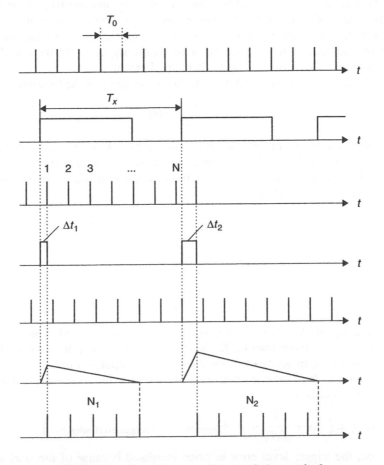

Figure 4.4 Time diagrams of interpolation method

greater number of intervals n accordingly. The first way requires a high-frequency generator and counter.

There is another conversion method which reduces the quantization error. It is the so-called interpolation method, in which instead of an integer number of reference frequency periods filling out the converted time interval, fractional parts of this period between the reference and the first counting pulse as well as the last counting and reference pulse are also taken into account [117]. This conversion method is illustrated in Figure 4.4.

The first counting pulse during the T_x, is delayed relative to the period's wavefront for time Δt_1, and the tail of T_x and the next counting pulse appearing after the tail–for time Δt_2. If it is possible to take precisely into account intervals Δt_1 and Δt_2, the quantization error would be excluded. The task of measuring the intervals Δt_1 and Δt_2 can be solved in the following way.

During the time interval Δt_1 a capacitor is linearly charged, and then discharged slowly 1000 times. This interval is filled by the same counting pulses and the number N_1 accumulates. The time interval Δt_2 is measured in the same way. As a result, the required time interval T_x is measured with absolute quantization error $T_0' = T_0/10^3$, which is equivalent to the filling by the counting pulses with a frequency 10^3 times more than f_0.

The speed of the indirect counting method is determined by the time interval T_x and the latency of the new measuring cycle. The latency can be reduced by two times in the case of the pulse signal or up to zero by using two counters working alternatively. The indirect counting method is a method with a non-redundant conversion time. However, its main fault is a high quantization error in the medium and high frequency range. So, for example, for conversion of a frequency $f_x = 10$ kHz and reference frequency $f_0 = 1$ MHz the quantization error will be 1%.

4.3 Combined Counting Method

Modifications to classical frequency-to-code conversion methods are possible using the combined conversion method and altering the frequency range. In the middle and high frequency range the conversion is carried out according to the standard counting method and in the low and infralow frequency range according to the indirect counting method [118].

The boundary frequency of adaptation at which methods are switched, is determined by the condition of maximum quantization errors (Equation 4.6) and (Equation 4.27) equality at $n = 1$ and determined as

$$f_{x\text{bound}} = \sqrt{\frac{f_0}{T_0}} \qquad (4.33)$$

Such a combination results in a number of properties, which expands the application of the combined method (Figure 4.5). The quantization error is essentially reduced at the ends of the conversion range of frequencies. The graph of $\delta_q(f_x)$ against the frequency has one global maximum at point $f_{x\text{bound}}$. Graphs of $\delta_q(f_x, T_0)$ and $\delta_q(f_x, f_0)$ are shown in Figures 4.6 and 4.7 respectively. With the increase of T_0 the quantization error δ_q is be decreased, and $f_{x\text{bound}}$ is shifted into the low frequencies. In turn, with

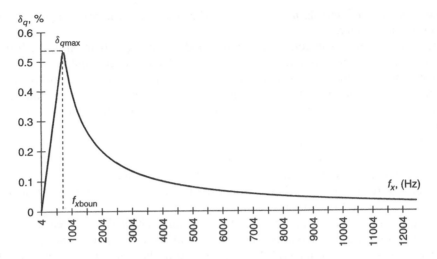

Figure 4.5 Graph of $\delta_q(f_x)$ for combined conversion method at $f_0 = 133333.33$ Hz and $T_0 = 0.25$ s

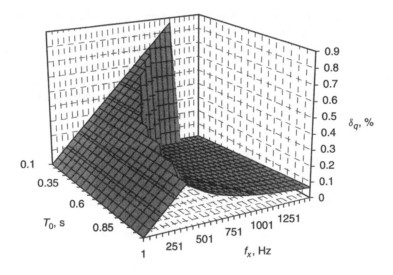

Figure 4.6 Graph of $\delta_q(f_x, T_0)$ for combined conversion method

the increase of f_0 the boundary switching frequency of conversion methods is shifted into the high frequencies.

A possible hardware configuration for the combined conversion method is shown in Figure 4.8 [96]. This circuit can be easily realized using silicon or FPGA.

The converter works as follows. Let trigger T be in the state corresponding to the frequency conversion mode. Simultaneously this trigger opens the logic element AND$_1$ and closes AND$_2$. Thus, each pulse of the input frequency sets up the counter CT2 to '0'. As the capacity of CT2 corresponds to the period duration at which the switching into a period conversion mode occurs, and the converted signal is in the high frequency

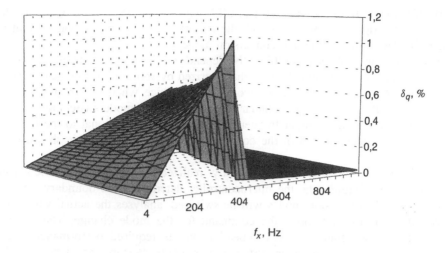

Figure 4.7 Graph of $\delta_q(f_x, f_0)$ for combined conversion method

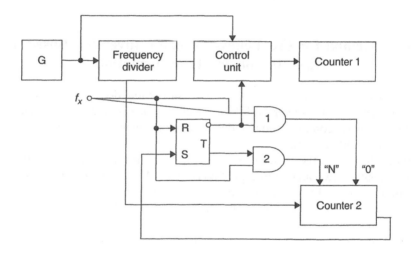

Figure 4.8 Frequency-to-code converted with adaptation of conversion modes

range, in each period of an input signal the reset of this counter to '0' will be earlier, than its overflow by reference frequency pulses from the frequency divider.

Owing to the absence of an overflow pulse for the counter CT2 the trigger will not change its state and the device will be constantly in the frequency conversion mode.

If the frequency of the converted signal decreases, the counter CT2 overflow during one period T_x will occur before the reset. In this case the trigger toggles to '1' by the overflow pulse of the counter CT2 and the device will be switched into the period conversion mode. Thus, the control unit allows the propagation of reference frequency pulses to the counter CT2 during one or several periods T_x. When the trigger T is switched to '1', the logical element AND$_1$ is closed and AND$_2$ opened. The next pulse of the input frequency will set up the counter CT2 to a state 'N' and the trigger to '0'. If the input frequency is such that CT2 overflow occurs in each period T_x, then each time the trigger is toggled to '1', the counter to 'N', and the control unit will switch the device into the period conversion mode.

That the counter CT2 is set to '0' or 'N' depends on the result of the current conversion cycle, and the mode switching is characterized by some hysteresis determined by the number N. It removes unstable work of the device when the frequency of the converted signal is close to the boundary frequency of the mode switching f_{xbound}.

Thus, depending on the input frequency, the number proportional to the frequency or the period is accumulated in the counter CT1. The trigger state is shown on the conversion mode.

The described converter may be easily realized based on a microcontroller. According to the initial data, the microcontroller determines the boundary frequency f_{xbound} at which conversion modes will be switched, analyses the actual value of the converted parameter and forms the command for the mode change. The particular choice of the microcontroller will depend upon the required performance and the operating range as well as other system requirements. Performance factors such as the accuracy and measurement range depend upon available on-board peripherals and the operating speed of the microcontroller.

4.4   Method for Frequency-to-Code Conversion Based on Discrete Fourier Transformation

Although this method for frequency-to-code conversion does not concern discrete conversion methods, but rather DSP-based methods of discrete spectral analysis, its application today is economic enough, due to rather inexpensive DSP microcontrollers, to be used, in smart sensors.

The block diagram of the device using this conversion method is shown in Figure 4.9. The device contains two analog-to-digital converters (ADC), two blocks of the discrete Fourier transform (DFT), two blocks for searching for the number of the maximum spectral component SC_{max} and the arithmetic logic unit (ALU) for calculation of the converted frequency.

The method can be successfully realized on the basis of the DSP microcontroller, including two ADCs or a single ADC with multiplexed input channels, for example, TMS320C24x families from Texas Instruments [119]. This DSP microcontroller

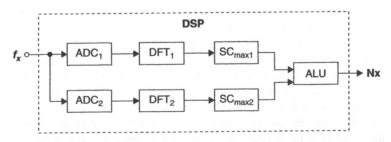

Figure 4.9   Device for frequency-to-code conversion based on method of discrete spectral analysis

includes two 10-bit ADCs, and also a set of other peripherals, useful which are in smart sensors: timers, PWM and CAN modules, etc.

The method of frequency-to-code conversion described in [120] consists of the simultaneous digitization of the input signal with drawing on N and $N-1$ points, the determination of the number of the maximum spectral component received with the help of the discrete Fourier transform and calculation using these numbers of the converted frequency.

The device works as follows. An input signal is descritized in the analog-to-digital converters, and discrete values N and $N-1$ are received accordingly. These values enter blocks DFT$_1$ and DFT$_2$ and after calculation of the discrete Fourier transform, the discrete spectrum is produced. Numbers of maximal spectral components are determined in blocks SC$_{max1}$ and SC$_{max2}$ accordingly. Then according to these numbers the ALU calculates the frequency.

The equation for the frequency determination on each spectrum can be given as

$$f_x = (K_1 + \ell_1 \cdot N) \cdot \Delta f \text{ and } f_x = K_2 + \ell_2 \cdot (N-1) \cdot \Delta f \tag{4.34}$$

Passing to the relative frequency $F = f_x/\Delta f$, we have:

$$F = K_1 + \ell_1 \cdot N \tag{4.35}$$

$$F = K_2 + \ell_2 \cdot (N-1), \tag{4.36}$$

where ℓ_1 and ℓ_2 are the number of periods of discrete spectrums for N and $N-1$ points, respectively.

Since $N > N-1$, and $\ell_1 \leq \ell_2$, then $\ell_2 = \ell_1 + X$, where $X = 0, 1, \ldots$, substituting ℓ_2 into Equation (4.36), we have $F = K_2 + (\ell_2 + X) \cdot (N-1)$. Let us express ℓ_1 as

$$l_1 = \frac{F - K_2 - X \cdot (N-1)}{N-1} \tag{4.37}$$

substituting this expression in Equation (4.35)

$$F = K_1 + \frac{F - K_2 - X \cdot (N-1)}{N-1} \cdot N \tag{4.38}$$

Multiplying both parts of the equation to $(N-1)$, we have

$$(N-1) \cdot F = K_1 \cdot (N-1) + F \cdot N - K_2 \cdot N - X \cdot (N-1)N. \tag{4.39}$$

Then removing the brackets we can write F as

$$F = K_2 \cdot N - K_1 \cdot (N-1) + X \cdot N \cdot (N-1).$$

By designating $K_y = K_2 N - K_1(N-1)$ and $N_y = N(N-1)$, we have $F = K_y + N_y X$, where $X = 0, 1, \ldots$, or

$$f_x = (K_y + N_y \cdot X) \cdot \Delta f \tag{4.41}$$

The range of the unequivocal frequency determination by this device is determined according to the equation

$$D_y = N_y \cdot \Delta f = N \cdot (N-1). \tag{4.42}$$

The frequency of the input signal can be determined according to the following equation:

$$f_x = \Delta f \begin{cases} K_2 \cdot N - K_1 \cdot (N-1), & \text{if } K_2 \geq K_1, \\ K_2 \cdot N - K_1 \cdot (N-1) + N \cdot (N-1), & \text{if } K_2 < K_1, \end{cases} \tag{4.43}$$

where $\Delta f = 1/T$ is the spectrum resolution; T is the considered interval for the input signal; N is the number of discrete points of DFT.

Let the input signal be a harmonious oscillation with frequency f_x. The spectrum of such a signal is shown in Figure 4.10(a). After digitization of the input signal and superimposing on the dot discrete spectrum (Figure 4.10(b) and (c) respectively) it is impossible to determine unequivocally the frequency of the input signal on each of the received spectrums separately, but comparing them, we find that the true frequency is at the point where the maxima of the periodic spectrums coincide.

Let us carry out an analysis using the criterion which reflects the relative range extension of the unequivocal frequency determination

$$n = \frac{D_y}{D_n}, \tag{4.44}$$

where n is the relative range extension; D_n and D_y are the ranges of the unequivocal frequency determination for the old and new devices respectively.

$$D_n = N \cdot \Delta f; \; D_y = N \cdot (N-1); \tag{4.45}$$

$$n = \frac{D_y}{D_n} = N - 1. \tag{4.46}$$

This method allows the range of the unequivocal frequency determination to be extended, and it is possible to achieve the necessary ratio between the range and the required descritization of the frequency determination.

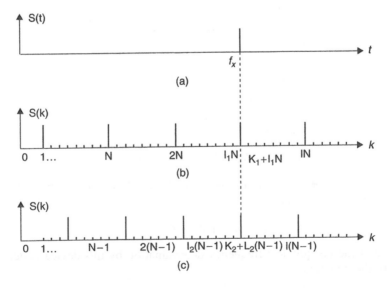

Figure 4.10 Spectrum of harmonious oscillation with frequency f_x

4.5   Methods for Phase-Shift-to-Code Conversion

The phase shift (φ_x)-to-code conversion can be reduced to the conversion of the time interval t_x on which two periodic sequences of pulses with the period T_x [117] are shifted.

The essence of the classical widespread method of the phase-shift-to-code conversion consists of the following. Sinusoidal voltages V_1 and V_2, the phase shift between which it is necessary to measure, are converted into short unipolar pulses (Figure 4.11). The strobing pulse t_x is formed from the first pair of pulses 1 and 2. It is filled out by the reference frequency pulses f_0. The number of pulses coming into the counter during the interval t_x:

$$n = f_0 \cdot t_x \tag{4.47}$$

A strobing pulse, equal to the period of the converted sinusoidal voltage, is formed in parallel. This pulse is also filled out by the reference frequency pulses f_0. The number of counted pulses during the period T_x is:

$$N = f_0 \cdot T_x \tag{4.48}$$

The phase shift is calculated according to the following equation

$$\varphi = \frac{360 \cdot n}{N} \tag{4.49}$$

The microcontroller core embedded in a smart sensor allows the choice of period necessary to determine the phase shift as well as observing phase shift fluctuations. There is also the possibility to convert average phase shifts during q periods.

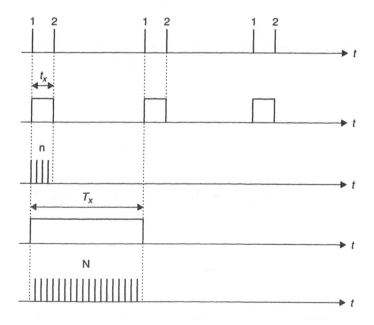

Figure 4.11   Time diagrams of method for phase-shift-to-code conversion

However, by using this method for the phase-shift-to-code conversion, high accuracy can be achieved only in low and infralow frequency ranges. In order to increase the conversion resolution and accuracy, the interpolation method, a method based on multiplication by f_x of the time interval proportional to the converted shift, or multiplication of the reference frequency f_0 by f_x can be used.

The basic components of the total conversion error of phase-shift-to-code converters, based on conversion of the phase shift to the time interval are:

- the noise influence on the conversion process

- the quantization error.

If the form of the converted signal differs from the rectangular, for example, sinusoidal, there are additional components of the conversion error:

- the trigger error

- the error caused by the difference of the form of the converted signals from the sine wave because of non-linear distortions.

Also, it is common in phase shift conversion that instrument error caused by non-identical channels has an effect. This error can be automatically corrected. Furthermore, microelectronic technologies for modern one-chip smart sensors practically guarantee the channel identity.

The standard deviation of the quantization error distributed according to Simpson's distribution law, is equal to:

$$\sigma_q = \frac{\Delta\varphi_q}{\sqrt{6}} \tag{4.50}$$

At $T_0 = $ const the quantization error increases with frequency f_x. By the given absolute conversion error $\Delta\varphi_q$ the higher limiting frequency is

$$f_{x\,max} = \frac{\Delta\varphi_q}{360 \cdot T_0} \tag{4.51}$$

Example: At $\Delta\varphi_q = 0.1^0$ and $T_0 = 10^{-6}$ ($f_0 = 1$ MHz) the greatest possible frequency of the converted signal should not exceed $f_{x\,max} = 277.77$ Hz. The low frequency is not limited.

Another classical method of the phase-shift-to-code conversion consists in the phase shift conversion for some periods during the constant conversion time T_{cycle}. Thus, the high converted frequency is considerably extended; however, the conversion time will be increased.

Summary

The choice of interface and the conversion technique depends on the desired resolution and the data acquisition rate. For maximum data acquisition rate, period measurement techniques (indirect counting techniques) can be used. The period conversion requires the use of a fast reference clock with available resolution directly related to the reference-clock rate.

Maximum resolution and accuracy may be obtained using the frequency-measurement (standard-counting), the pulse-accumulation or integration techniques. Frequency measurements provide the added benefit of averaging out the random- or high-frequency variation (jitter) resulting from noise in the signal.

The comparative simplicity of the standard direct counting method as well as the indirect counting method in comparison with other methods (for example, methods of tracing balance, requiring the use of closed structures), high performance and universality have contributed to their popularity. They have become the basic, traditional, classical methods despite some restrictions and faults.

DSP-based methods for the frequency-to-code conversion of a harmonious signal based on the discrete spectral analysis have been developed due to advances in microelectronics. Today their application is economic enough, due to rather inexpensive DSP microcontrollers, to be used in smart sensors.

Further development of microelectronic technologies, microsystems and smart sensors, and new advanced methods of frequency (period)-to-code conversion aim to:

- increase accuracy, speed and metrological reliability of the measurement of absolute and relative values, and also their ratio

- expand functionality and the conversion range

- automatize completely all procedures of measurement, control, digital processing, parametric adaptation and self-diagnostics

- simplify the circuitry or minimize a chip area

- reduce the cost, weight, dimensions, power consumption, etc.

5

ADVANCED AND SELF-ADAPTING METHODS OF FREQUENCY-TO-CODE CONVERSION

In spite of the fact that today frequency can be measured by the most precise methods by comparison with other physical quantities, precise frequency-to-code conversion with the constant quantization error in a wide specified measuring frequency range (from 0.01 Hz up to some MHz) and with non-redundant conversion time can only be realized based on novel measurement methods for frequency-time parameters of the electric signal. This requires additional hardware costs and arithmetic operations: multiplication and division for calculations of the final result of conversion. Therefore, additional measuring (conversion) devices should be included in the microsystem. These include two or more multidigital binary counters, the multiplier and the code divider, logic elements, etc. Different design approaches are used. In the authors' opinion, a successful solution is the use of a microcontroller core in such microsystems. In this case, with the aim to minimize the built-in hardware, it is expedient to take advantage of the program-oriented methods of measurement developed for frequency-time parameters of signals.

5.1 Ratiometric Counting Method

We shall first consider the idea of the original method of the discrete count [98,117], called the *ratiometric counting method*, which allows frequency-to-code conversion with a small constant error in a wide frequency range, we shall then consider how this method can be realized.

Let's assume that the converted periodic signal is in the form of sine wave. By means of the input forming device it will be transformed into a periodic sequence of pulses, the period T_x of which is equal to the period of the converted signal. There are various devices and principles for the transformation of periodic continuous signals into a sequence of rectangular pulses. For the practical usage, we provide a simple enough circuit of such an input shaper (Figure 5.1). It has an input voltage range from

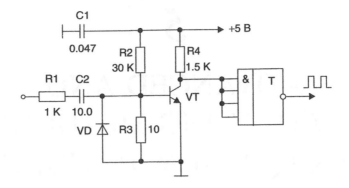

Figure 5.1   Circuit of input forming device

$+0.035$ up to $+24$ V. The device includes an input amplitude limiter, an amplifier and a Schmitt trigger.

Regardless of this sequence the first reference time interval (gate time) T_{01} is formed (Figure 5.2). It is filled out by N_1 pulses of the periodic sequence. The number N_1 is accumulated in the first counter. The converted frequency f'_x is determined according to the following

$$f'_x = \frac{N_1}{T_{01}} \tag{5.1}$$

The frequency deviation from the value f_x is determined by the quantization error, the reduction of which is the aim of this method.

Simultaneously, the second gate time T_{02}, is formed. Its wavefront coincides with the pulse, appearing right after the start of the first gate time T_{01}, and the wavetail, with the pulse appearing right after the end of the first gate time T_{01}. Thus, the duration of the second gate time T_{02} is precisely equal to the integer of the periods of the converted signal, i.e.

$$T_{02} = N_1 \cdot T_x \tag{5.2}$$

The wavefront and the wavetail of the formed gate time are synchronized with the pulses of the periodic input sequence generated from the input signal, therefore the rounding error is excluded. The second gate time is filled out by pulses of reference frequency f_0, whose number is accumulated in the second counter.

The formula for the calculation of the converted frequency can be obtained in the following way. The number of pulses which have got into the second gate time, as can be seen from Figure 5.2, is determined by the ratio

$$N_2 = N_1 \cdot T_x/T_0 = N_1 \cdot f_0/f_x, \tag{5.3}$$

hence

$$f_x = \frac{N_1}{N_2} \cdot f_0, \tag{5.4}$$

where f_0 is the reference frequency. This can be done in a PC or a microcontroller core along with the offsetting and scaling that must often be performed.

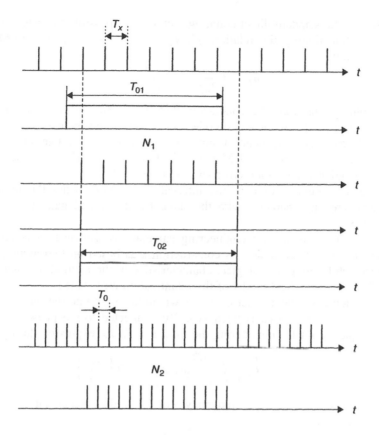

Figure 5.2 Time diagrams of ratiometric counting method

The accuracy of the frequency measurement is determined by the quantization error of the time interval $N_1 T_x$.

Let's withdraw the equation for the relative quantization error δ_q for the frequency-to-code conversion. First of all we shall determine the maximum value of the relative quantization error for the time interval $T_{02} = N_1 T_x$. As this interval is filled out by counting pulses with the period T_0, the maximum absolute error is $\Delta_2 = \pm T_0$, and the maximum relative error is:

$$\delta_2 = \pm \frac{T_0}{T_{02}} = \pm \frac{T_0}{N_1 \cdot T_x} \tag{5.5}$$

The equality $N_1 T_x = T_{02}$ can be presented as $f_x = N_1 / T_{02}$. Then according to the rules of the error calculation for indirect measurements the measurement error of the function f_x is connected with the measurement error of the argument T_{02} by the ratio (with the accuracy of the second order of smallness) $\delta_q = \delta_2$. After substitution of δ_2 from Equation (5.5) we have:

$$\delta_q = \pm \frac{T_0}{N_1 \cdot T_x} = \pm \frac{f_x}{N_1} \cdot T_0 \tag{5.6}$$

According to the standard direct counting method it is possible to write the equality $T_{01} = N_1/f_x'$. Substituting the relation $f_x'/N_1 = 1/T_{01}$, into Equation (5.6) instead f_x/N_1 we obtain:

$$\delta_q = \pm\frac{T_0}{T_{01}} = \pm\frac{1}{f_0 \cdot T_{01}} \tag{5.7}$$

This formula let us draw the conclusion that the maximum value of the relative quantization error for the frequency-to-code conversion for this method does not depend on converted frequencies and, hence, is constant in all conversion ranges of frequencies.

For a reference frequency $f_0 = 1$ MHz and the first gate time $T_{01} = 1$ s the maximum value of the relative quantization error will be $\delta_q = \pm10^{-4}\%$.

If, by the measurement of the time interval $T_{02} = N_1 T_x$ using the interpolation method considered in Chapter 4 with the same frequency and gate time we obtain $\delta_q = \pm10^{-7}\%$.

Equation (5.7) is suitable for engineering calculations. However, this equation is a simplified mathematical model. Its construction was based on two assumptions. In order to research limiting metrological characteristics of the method it is necessary to use the refined mathematical model of the quantization error [118].

In the common case, the number of pulses which have got into the second reference time interval T_{02} (Figure 5.2), is unequivocally connected to the phase crowding of reference frequency fluctuations during the time $\tau = N_1 T_x$, expressed through 2π:

$$N_2 = \text{ent}\left\{\frac{1}{2\pi}\int_{t-\tau/2}^{t+\tau/2}\omega_o\,dt\right\} = \text{ent}\left\{\frac{\omega_o\tau}{2\pi}\right\}, \tag{5.8}$$

where ω_0 is the radian reference frequency; ent $\{\ \}$ is the integer part of a number. Omitting the ent, we have:

$$N_2 = \frac{N_1 \cdot T_x}{T_0} \tag{5.9}$$

On the other hand, on the strength of formula for T_{02}, it is possible to write down

$$\delta_q = \pm\frac{f_x}{(N_1 + 1) \cdot f_o} = \pm\frac{T_0}{T_{01}} = \pm\frac{1}{N_2} \tag{5.10}$$

This formula represents the refined mathematical model for the quantization error and can be used for frequency-to-code converter simulation for studying limiting metrological characteristics.

The graph of δ_q, T_{02} against f_x is shown in Figure 5.3 and the functional relationship $\delta_q(f_x, T_{01})$ in Figure 5.4. Functions $\delta_q = f(f_x)$, $\delta_q(f_x, T_{01})$ and $T_{02} = \phi(f_x)$ have the finite discontinuity of the first kind. As the modelling results show, the quantization error in the neighbourhood of the point $f_{x\,min}$ is two times less than that calculated according to Equation (5.7). Thus, the second reference time interval T_{02} is increased by as many times.

Finally, we consider the block diagram of the converter (Figure 5.5), which realizes the frequency-to-code conversion according to the considered ratiometric counting method.

It contains two counters and a D-trigger clocked with the sensor output. All counter functions can be provided by the counter/timer peripheral component, which interfaces to many popular microcontrollers.

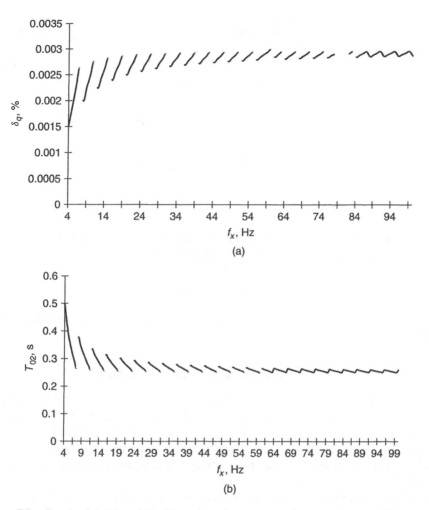

Figure 5.3 Graph of $\delta_q(a)$ and $T_{02}(b)$ against f_x at $T_{01} = 0.25$ s and $f_0 = 133\,333.33$ Hz

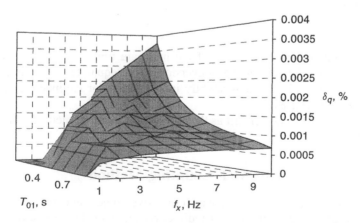

Figure 5.4 Graph of δ_q against f_x and T_{01} at $f_0 = 133\,333.33$ Hz

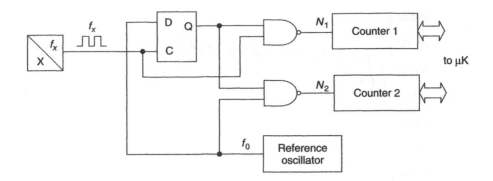

Figure 5.5   Ratiometric simplified counting scheme

This method can also be used for the period (T_x)-to-code conversion. Thus, the period is calculated according to the following formula

$$T_x = \frac{N_2}{N_1} \cdot T_0 \qquad (5.11)$$

The main demerit of the considered conversion method is the redundant conversion time.

5.2   Reciprocal Counting Method

A variation of the method described above is the reciprocal counting method (Figure 5.6). The cyclicity of the conversion T_{cycle} is determined by the reset pulse. The beginning of the counting interval T_{count} coincides with the next pulse of sequence f_x, appearing after the ending of the 'Start' pulse, and the end coincides with the next pulse f_x, appearing after the 'End' pulse [105]. The frequency is calculated during the interval T_{calc}.

The frequency or the period are calculated similarly as described above according to Equations (5.4) and (5.11) accordingly. The quantization error is calculated according to the following:

$$\delta_q = \pm \frac{1}{f_0 \cdot T_{count}} \qquad (5.12)$$

However, in this method, the value of the first gate time T_{count} can be determined only approximately. Having set the conversion cycle equal to 1 s, it is possible to assume only that $T_{count} \cong 1$ s. It is slightly inconvenient for engineering calculations of the quantization error δ_q.

This method has the same demerit as the ratiometric counting method–the redundant conversion time.

5.3   M/T Counting Method

The so-called M/T counting method, which also overcomes demerits of conventional methods (standard direct counting and indirect counting methods) and achieves high resolution and accuracy for a short detecting time was described in [110].

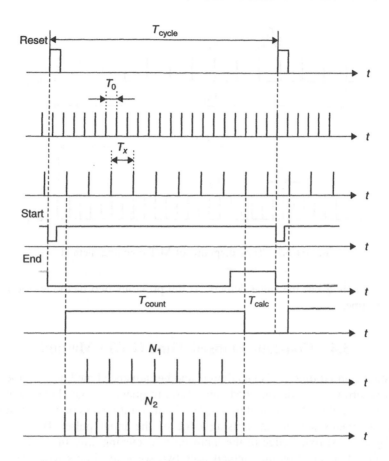

Figure 5.6 Time diagrams of reciprocal counting method

Time diagrams of the M/T counting method are shown in Figure 5.7. The detecting time T_{02} is determined by synchronizing the generated output pulse after a prescribed period of time T_{01}.

It is easy to notice that the M/T method differs from the above described ratiometric counting method by the synchronization of the first reference time interval T_{01} with the pulse of the converted frequency f_x. The converted frequency or the period are determined similarly according to Equations (5.4) and (5.11). The quantization error does not depend on the converted frequency and is constant for all frequency ranges.

The detecting time is calculated by the following formula:

$$T_{02} = T_{01} + \Delta T \tag{5.13}$$

The resolution Q_N is calculated when the clock pulses N_2 are changed by one:

$$Q_N = f_0 \cdot N_1 \left(\frac{1}{N_2 - 1} - \frac{1}{N_2} \right) = \frac{f_0 \cdot N_1}{N_2 (N_2 - 1)} \tag{5.14}$$

The method uses three hardware timers/counters. One of them works in the timer mode in order to form the first time interval T_{01}, the rest are used in the counter mode.

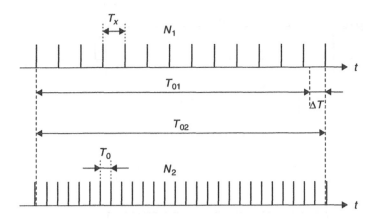

Figure 5.7   Time diagrams of M/T counting method

The M/T method has the same demerit as the previous two methods–the redundant conversion time.

5.4   Constant Elapsed Time (CET) Method

The constant elapsed time (CET) method is another advanced method with the constant quantization error in a wide specified conversion frequency range [111]. Like the M/T counting method, the CET method is based on both counting and period measurements. Both measurements starting at a rising edge of the input pulse. The counters are stopped by the first rising edge of the input pulse occurring after the constant elapsed time T_{01}. For its realization the method uses two software timers with two hardware timers/counters inside the microcontroller.

In order to eliminate repetitive starting and stopping counters that require a reinitialization, the time counters run continuously and the pulse counter is read at the end of each time interval. This method has the same demerit: the redundant conversion time.

5.5   Single- and Double-Buffered Methods

The single-buffered (SB) method is based on both pulse counting and measurement of the fractional pulse period before the interrupt sample time [112]. Instead of repetitive enabling/disabling of the capture register function, this function is always enabled. Each rising edge of the synchronized input pulse f_x stores the content of the timer in the timer capture register. The same pulse is the clock for the pulse counter. The interrupt requests are generated using the interval timer, without any link to the pulse rising edges.

The time difference N_2 in this method is determined using the difference between two readings of the capture register during the current and previous interrupt service routine accordingly.

The pulse difference N_1 is determined using the difference between two readings of the pulse counter during the current and previous interrupt service routine accordingly.

The frequency f_x is determined using the pulse difference, the time difference and the clock frequency f_0 of the timer as Equation (5.4).

Each frequency-to-code converter consists of a free running timer with a timer capture register, a pulse counter and an interrupt generator. The hardware complexity for such a system, using a software timer instead of an interval timer, is similar to that of the CET method. Hardware for the SB system is available in some microcontrollers.

The SB method has the following disadvantages [112]:

1. The reading of the timer capture register can be erroneous, if the reading is performed during the store operation of the content of the free running timer. The other problem occurs if the reading of the pulse counter is done during counting. Both problems require synchronization hardware.

2. The rising edge of an external pulse at the time interval between the reading of the timer capture register and the reading of the pulse counter will cause the inadequacy of the content of the pulse counter compared to the necessary content. This is a source of a numerical error, especially during the low frequency measurement.

The converter based on the double-buffered (DB) method consists of an interval timer with an associated modulus register, a timer capture register, an additional capture register and a pulse counter with the counter capture register.

The problems of the SB method are inherently solved using synchronization with the rising and falling edges of the system clock. The maximum measured frequency for the DB method is limited by the synchronization logic. The equivalent hardware complexity for both methods is similar, because of the free running the timer consists of the register and associated logic. The maximum measured frequency for the DB method is, however, not limited by the software loop only by the hardware.

The measurement error for the SB and DB methods is caused by the synchronization of the pulse signal. The worst-case error is caused by missing one count of the system clock f_0. The lower error limit of the SB and DB methods is the same as the error limit of the ratiometric, reciprocal, M/T and CET methods, while the higher error limit of the SB and DB methods is twice as much as the error limit of these methods. Like all the mentioned methods SB and DB methods also have the redundant conversion time in all specified conversion frequency ranges except for the nominal frequency.

5.6 DMA Transfer Method

The direct memory access (DMA) method [113] provides an average frequency measurement of the input pulses, based on both pulse counting and time measuring for constant sampling time. The time is measured by counting pulses of the reference clock with the frequency f_0 in a free-running timer. Each rising edge of the input pulse activates a DMA request. The DMA controller transfers the content of the free-running timer into the memory (analogous to the timer capture register in the SB method) and decrements the DMA transfer counter (analogous to the pulse counter in the SB method). After each constant sampling time, the interval timer generates an interrupt request to the microcontroller, which reads the contents of both the DMA transfer counter and the memory in an interrupt routine. The measured frequency is calculated as in Equation (5.4).

Due to some specific errors for the DMA method the maximum error is greater than the error of all the above advanced conversion methods (sometime more than 10 times greater) in the case of the coincidence between an input pulse and the execution of the longest microcontroller instruction, just prior to reading in an interrupt routine.

5.7 Method of Dependent Count

The method of the dependent count (MDC) was proposed in 1980 [116]. It combines the advantages of the classical methods as well as the advanced methods ensuring a constant relative quantization error in a broad frequency range and at high speed. The method was developed for the measurement of absolute [116], relative frequencies [121], periods, their ratios, difference and deviations from specified values. The creation of these methods was a logical completion of the goal-oriented search for optimal self-adopted algorithms developed earlier by the authors.

One of the essential advantages of this method is the possibility to measure the frequency $f_x \geq f_0$. In this connection we shall use the following denotations for the deducing the main mathematical formulas: F is the greater of the two frequencies f_x and f_0; f is the lower of the two frequencies f_x and f_0.

In general, the method of the dependent count (Figure 5.8) is based on simultaneous:

- separate counts of the periods of two impulse sequences of measurand frequencies and reference frequencies at the relative and absolute measurement method accordingly

- comparisons of the accumulative number with the number N_δ, program-specified by the relative error δ at the frequency measurement, or at first with the number N_1 determined by the error δ_1 of identification of the greater of the two frequencies and only then with the number N (at measurement of the ratio)

- forming of the reference gate (quantization window) T_q equal to the integer number N_x of the periods τ of the lower frequency f

- quantization of the created reference time T_q by duration of the periods T of the greater frequency F, the number of which is $N \geq N_\delta$.

The microcontroller reads out the number N (for the consequent summation with the number N_δ and calculation of the result of the measurement) at the moment of appearance of the impulse of the next period τ, terminating the impulse count.

In methods of the dependent count, the frequency-to-code conversions are supplemented by arithmetic calculations. The latter can be executed simultaneously with the consequent measurement or before its beginning. The methods are suitable for single channel as well as for multichannel synchronous frequency conversions.

Let's consider the algorithms of absolute and relative methods of conversion in more detail.

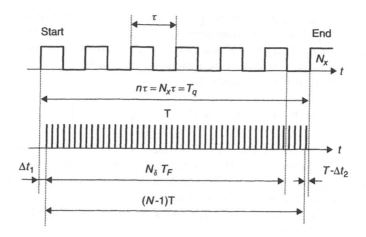

Figure 5.8   Time diagrams of the absolute method of dependent count

5.7.1   Method of conversion for absolute values

The measurement (conversion) of frequency and its deviations from the program-specified values is executed automatically during one time step T_q. Modes of one-time, cyclic or continuous measurements with high accuracy and speed are possible. The measurements are realized by the separate count of the impulse with the normalized width of two sequences with the frequencies f_0 and f_x, summation and the continuous comparison of the stored numbers with the specified number N_δ for determination of the moment of equality to one of them. Simultaneously, the quantization window T_q is formed, equal to the integer number of the periods τ of the lower frequency f (f_x or f_0). This interval is quantized by duration of the periods T of the greater frequency (f_0 or f_x). The accumulation of the number N in the frequency F counter is necessary, but it is an insufficient condition for the count termination and summation of periods of both sequences. The impulse count is stopped only at the moment of appearance of the wavefront of the consequent period, ensuring the multiplicity of T_q to duration of the period τ. After reading the numbers $n = N_x$ and N, fixed in the counters of lower f and greater F frequencies accordingly, the measuring conversion is supplemented by the numerical according to the formulas:

$$f = F\frac{n}{N} = F\frac{N_x}{N_0} = F\frac{N_x}{N_\delta + \Delta N}, \tau = \frac{1}{f} = T\frac{N_\delta + \Delta N}{N_x}, \qquad (5.15)$$

where $N = N_\delta + \Delta N$ is the total number of periods T, that was accumulated in the counter of the frequency F during the quantization window T_q; ΔN is the complement number to the specified number N_δ of the periods T that will be accumulated in the counter of the frequency F after its nulling up to the measuring termination.

The time diagram of the count of the periods τ of the lower frequency f and the periods T of the greater frequency at the frequency measurement f is shown in Figure 5.8. It is obvious that:

$$T_q = n \cdot \tau = N_x \cdot T_x = N \cdot T + \Delta t_1 - \Delta t_2 = N \cdot T \pm \Delta_q, \qquad (5.16)$$

where $\Delta_q = \Delta t_1 - \Delta t_2$ is the absolute quantization error (the error of the method) of the time interval T_q. This error arises because of the non-synchronization of the first and last impulse of the frequency F with wavefronts of impulses of the frequency f, determining the start and the end of the quantization window.

The time of the measurement (conversion) is:

$$T_q = N \cdot T = (N_\delta + \Delta N) \cdot T = \frac{1}{\delta}\left(1 + \frac{\Delta N}{N_\delta}\right) \cdot T \qquad (5.17)$$

with the error, not exceeding the $\pm T$. This time does not depend on the measurand frequency in the frequency range D_1. The range D_1 can vary from infralow frequencies up to the frequency $f_0 = f_{max}$. The latter is determined by the greatest possible frequency that can be counted by the timer/counter.

In case, when $\tau = T_x$ (mode $f_0 \geq f_x$) and $T = T_0 = 1/f_0$, Equations (5.15) and (5.17) should be written as:

$$f_x = f_0 \frac{N_x}{N_\delta + \Delta N}, \; T_x = T_0 \frac{N_\delta + \Delta N}{N_x}, \; T_q = \frac{T_0}{\delta}\left(1 + \frac{\Delta N}{N_\delta}\right). \qquad (5.18)$$

If $\tau = T_0$ and $T = T_x$ (mode $f_x > f_0$), then:

$$f_x = f_0 \frac{N_\delta + \Delta N}{N_x}, \; T_x = T_0 \frac{N_x}{N_\delta + \Delta N}, \; T_q = \frac{T_x}{\delta}\left(1 + \frac{\Delta N}{N_\delta}\right) \qquad (5.19)$$

The frequency range is extended up to $2D_1$ and overlapped only by one specified measuring (working) range.

5.7.2   Methods of conversion for relative values

Unlike the algorithm for measurement of absolute values, the algorithm for measurement of relative values is executed in two steps [122]. The measurement of the ratio f_{x1}/f_{x2} is executed without using the reference frequency f_0 (Figure 5.9).

The first step is used for the determination of the greater one of the two unknown frequencies. To this effect the following procedures are carried out:

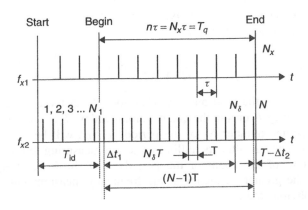

Figure 5.9   Time diagram of method for frequency ratio measurements

1. The calculation of the number $N_1 = 1/\delta_1$ according to the program-specified relative error of determination of the greater frequency.

2. The separate count.

3. The summation of $T_{x1} = 1/f_{x1}$ and $T_{x2} = 1/f_{x2}$ periods of impulse sequence.

4. The comparison of the obtained sums up to reaching the equality to the number N_1.

The impulse count of both frequencies is completed at the moment of accomplishment of this equality. The information about the greater of two frequencies is entered into the microcontroller.

The second step is used for measurement of the frequencies ratio f_{x1}/f_{x2}, if $f_{x1} \leq f_{x2}$, or f_{x2}/f_{x1}, if $f_{x1} \geq f_{x2}$. Thus:

1. The number $N_\delta = 1/\delta$ is calculated on the program-specified relative error δ of ratio measurement.

2. The impulses of both sequences are calculated.

3. Each of the accumulated sums is compared period by period with the number N_δ according to the algorithm of the absolute method of the dependent count.

The integer number n of the periods τ is fixed into the counter of the lower frequency $f(f_{x1}$ or $f_{x2})$. The count of impulses of the frequencies f_{x1} and f_{x2} is stopped after the counting by the counter of the frequency F (f_{x2} or f_{x1}) the number $(N_\delta + {}^\wedge N)$ of the periods T. The result of measurement is calculated according to the formula:

$$\frac{f}{F} = \frac{n}{N} = \frac{N_x}{N_\delta + \Delta N} \quad \text{or} \quad \frac{\tau}{T} = \frac{N_\delta + \Delta N}{N_x} \tag{5.20}$$

and displayed in the specific units of measurement (relative, %, per mille ‰, etc.). With regard for the result of determination of the greater of the two frequencies, the calculations will be carried out according to the formulas:

$$\frac{f_{x1}}{f_{x2}} = \frac{N_{x1}}{N_{x2} + \Delta N_2} \quad \text{if } f_{x1} \leq f_{x2}$$

and

$$\frac{f_{x2}}{f_{x1}} = \frac{N_{x2}}{N_{x1} + \Delta N_1} \quad \text{if } f_{x2} \leq f_{x1} \tag{5.21}$$

Thus, the relative error of measurement of the frequency ratio will not exceed the program-specified error δ.

Let us consider another method for frequency ratio. The second method is more complex. Its realization requires the addition generator of the reference frequency f_0, the third counter of impulses for the frequency f_0 and more complex software for the microcontroller.

Due to exception of the procedure for determination of the greater frequency (Figure 5.9), the ratio f/F is measured by one step. The measurements of frequencies f and F are executed simultaneously according to the method of the dependent count for absolute values, but with some features. In this case, the following procedures are executed:

- The separate and simultaneous count of normalized impulses of frequencies F, f and f_0.

- Forming of the reference time intervals T_{q1} and T_{q2} equal to the sum of the integer of periods T and τ.

- Their simultaneous and independent quantization by duration of the period T_0.

The number of the latter is accumulated and continuously compared with the beforehand given $N_\delta = 1/\delta$. The accumulation of the number N_δ in the counter of the frequency f_0 is necessary, but there is the insufficient condition for the count termination and summation of the durations of periods τ and T. The impulse count of the greater frequency F is terminated at first at the moment of appearance of the impulse followed by its period T. Due to this the equation $T_{q1} = N_{x1}T$ is true. Simultaneously the result of the frequency F to code conversion (the numbers N_{x1} and ΔN_1) is read out. Then the impulse count of the lesser frequency f is stopped. It happens at the moment of appearance of the impulse followed by its period τ. The last fact ensures the equality $T_{q2} = N_{x2}$. After that the count of periods T_0 with simultaneous reading of the numbers N_{x2} and ΔN_2, and determination of the greater of two frequencies f_{x1} or f_{x2} is stopped. The analog-to-digital conversion is supplemented by the calculations:

$$F = f_{x2} = f_0 \frac{N_{x1}}{N_\delta + \Delta N} = f_0 \frac{N_{x1}}{N_1};$$

$$f = f_{x1} = f_0 \frac{N_{x2}}{N_\delta + \Delta N} = f_0 \frac{N_{x2}}{N_2}; \qquad (5.22)$$

$$\frac{f}{F} = \frac{f_{x1}}{f_{x2}} = \frac{N_{x2}N_1}{N_{x1}N_2},$$

where ΔN_1 and ΔN_2 are the complementary numbers up to the number N_δ. They are read out from the counter of the reference frequency f_0 after the first and second interrupt. The ratio of periods T_{x1} and T_{x2} can be determined from Equations (5.22). Besides it is possible to determine the absolute and relative differences of two frequencies or periods; absolute and relative deviations of the value of controlled parameters from one or several program-specified values.

The circuitry of the first and second methods for frequency ratio conversion is rather simple. It does not require a lot of hardware and accepts their margin and the program choice of one of them. The example of such a universal converter for the ratio of two frequencies constructed with the most complex hardware is shown in Figure 5.10.

The microcontroller expands the possibilities of such converters and realizes the functions of transmission of the results of conversion of frequency deviation by the pulse-width or code-impulse signals. It is necessary for realization of digital intelligent distributed sensor systems, the multichannel converter, etc. The hardware–software realization of the method for frequency ratios expands the range of measurand frequencies, due to the use of the hardware and virtual counters, realized inside the microcontroller. Therefore, the range of measurand frequencies practically is unlimited even at low frequencies.

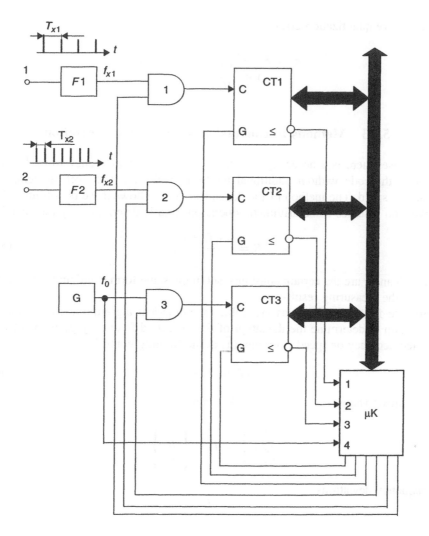

Figure 5.10 Frequency ratio f_{x1}/f_{x2} to code converter

The realization of wide-range frequency converters requires two addition hardware counter dividers of input frequency, some triggers and digital logical circuits.

For both methods of measurement the time of quantization T_q is non-redundant in all ranges of measurand frequencies and can be varied in the limits:

$$\frac{T}{\delta} \leq T_q = \frac{T}{\delta}\left(1 + \frac{\Delta N}{N_\delta}\right) \leq \frac{2T}{\delta}.$$
(5.23)

The absolute error of the methods is:

$$\Delta_q = \frac{\Delta t_2 - \Delta t_1}{T}.$$
(5.24)

The relative quantization error:

$$\delta_q = \delta \frac{1}{1 + \dfrac{\Delta N}{N_\delta}} \leq \delta_{q\,max} = \delta. \tag{5.25}$$

5.7.3   Methods of conversion for frequency deviation

In some cases there is a necessity to convert deviations of frequency from the preset value into the code without additional hardware and with high software controlled accuracy or speed. The method of the dependent count allows these possibilities. Such a conversion assumes some arithmetic operations and is determined by the equation

$$R = \frac{a \cdot f_x \pm b}{c}, \tag{5.26}$$

where a, b and c are the certain constants, set in the same format and units of measurements, as the measuring results.

Thus, the following formulas are valid for numerical algorithmic transformations with the aim to determine the deviation of the measured frequency f_x with software-controlled accuracy or speed from one (as at the limiting control)

$$\Delta f_x = f_x - b \tag{5.27}$$

or some preset values (as at the admissible control)

$$\left.\begin{array}{c} \Delta f_{x1} \\ \Delta f_{xi} \\ \Delta f_{xn} \end{array}\right\} = f_x - \left\{\begin{array}{c} B_1 \\ B_i \\ B_n \end{array}\right\} \tag{5.28}$$

and frequency deviations

$$\Delta f_{xd} = \frac{f_x - b}{f_{x\text{basic}}}. \tag{5.29}$$

Let's specify, that the task of the control for frequency-time parameters of signals can also be solved: at the limiting control it will be determined if $f_x <> f_{x\text{basis}}$, and at the admissible control, whether there is a controllable frequency inside some specified area of allowable values.

5.7.4   Universal method of dependent count

The algorithm of the universal method of the dependent count integrates algorithms of both absolute and relative methods. The distinctive feature of the measurement for absolute values according to the algorithm of the universal method of the dependent count is, that it will be executed in two steps. Owing to these, no additional measures for determination of the greater of two frequencies are necessary.

Thus, all three algorithms of the method include: the procedure of automatic count number of normalized on duration impulses, for assurance of the inequality $\tau \ll T_{x\,min}$;

the procedure of period-by-period comparison of the accumulated sums with the specified number; the procedure of processing of results of measurements. Theoretically, the count does not bring in an error by the usage of appropriate devices for its implementation.

The comparison error that is determined by the speed of the used devices and a number of bits of compared numbers can be also neglected. The calculation error of the result of measurement under the specified formulas can be neglected as well. The rounding error is distributed according to the uniform unbiased (equiprobable) distribution law. Therefore, the error of the frequency-to-code conversion will be determined only by the quantization error, the reference frequency stability and the trigger error in case of a harmonic signal and the effect of interference.

The consequent digital signal processing: scaling, determination of the difference and deviation of measurand values from program-specified, minimax, statistical and weight signal processing, sensor characteristics linearization, extrapolations, etc. can be used for the further processing of the result of the measurement.

5.7.5 Example of realization

The simplicity of the method realization using minimum hardware is an important advantage. The schematic diagram of the frequency-to-code converter based on the universal method of the dependent count is shown in Figure 5.11.

At the beginning the measuring mode (absolute or relative values) and the required relative errors δ_1 and δ are set. Then the microcontroller calculates the numbers N_1, N_δ and initializes the counters CT1, CT2. The measurements begin after actuation of the trigger T4 by a start impulse through the logical element OR_1. Then the elements AND_5, AND_6 are actuated. The formed impulses with the frequencies f_{x1} and f_{x2} at ratio measurement or f_{x1} and f_0 at frequency measurement (the trigger T3 is switched off) feed on the inputs of the decrement counters. The signal from the zeroized counter is fed on the interrupt input of the microcontroller. The microcontroller switches off the trigger T4 and terminates the impulse count of both frequencies.

If the frequency f_{x2} was greater at the ratio measurement (the trigger T3 is actuated), the microcontroller downloaded the number $N_\delta = 1/\delta$ only in the counter CT2 at repeated initialization. Then the trigger T1 is actuated again. The first impulse of the frequency f_{x1} through the open element AND_1 and the element OR_1 actuates the trigger T4 again and switches off the trigger T1. The new count cycle of periods and the decrement of the counters begins: CT1 from the $N_{max} = 2^i$ (where $i = 16, 32, 64$); CT2 from the N_δ. After zeroizing the last one, the signal from the inverse output CT2 goes on the microcontroller's input. The μK actuates the trigger T1 again. The consequent impulse of the frequency f_{x1} switches off the triggers T1, T4 and eliminates the count of impulses of both frequencies. The signal from the inverse output of T4 goes on the third microcontroller's input and eliminates the count at the ratio measurement f_{x1}/f_{x2}.

If f_{x1} is the greater frequency, the microcontroller downloads the number N_δ only in the counter CT1 at initialization. The decrement of the counters is executed in a similar way from the values N_δ and N_{max}. Thus, the quantization window is formed. It is equal to the integer number of the periods $T_{x2} = 1/f_{x2}$. After the last nulling, the signal from the inverse output of the trigger T4 goes again on the μK's input and eliminates the

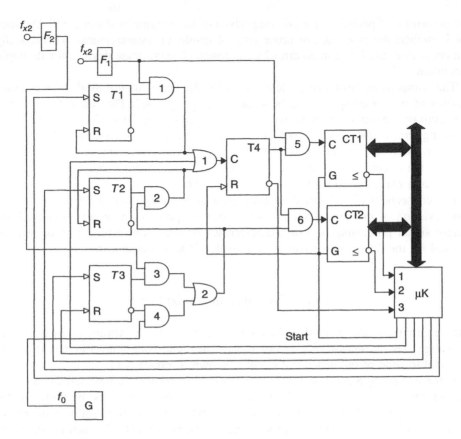

Figure 5.11   Schematic diagram of universal method of dependent count

measurement. The μK reads the numbers N_{x2} and ΔN_1 (or N_{x1} and ΔN_2, if $f_{x1} \leq f_{x2}$) and calculates the result of the measurement according to Equation (5.21).

The average value of frequency is measured similarly. The mode of determination of the greater frequency is also necessary in this case. Due to this, after completion of the second step of measurement with accounting the result on the first step, the μK calculates the result as:

$$f_x = f_0 \frac{N_x}{N_\delta + \Delta N}, \; T_x = T_0 \frac{N_\delta + \Delta N}{N_x}, \; \text{if } f_x \leq f_0, \tag{5.30}$$

$$f_x = f_0 \frac{N_\delta + \Delta N}{N_x}, \; T_x = T_0 \frac{N_x}{N_\delta + \Delta N}, \; \text{if } f_{x1} > f_0 \tag{5.31}$$

The determination of the greater frequency is possible without necessity of execution of the first step. It requires some more complicated software.

Let's note that this scheme is one of the most complex hardware realizations of the method, except the multichannel synchronous one with the simultaneous conversion of frequencies in all channels. It can be realized in the silicon, PLD or FPGA. The most simple is the program-oriented realization [123–124]. The combine implementation [125] is somewhat more complex. In these cases, the measuring channel will be realized completely or partially on the virtual level inside the functional-logical architecture of the computer power.

5.7.6 Metrological characteristics and capabilities

The accuracy and time of measurement are the main characteristics of various methods for frequency-to-code conversion. These characteristics are the main quality indexes at any automatic measurements. However, in the case of frequency measurements they are more interconnected. The method of the dependent count differs from other conventional methods in that the indicated correlation is very loosed, because the quantization window T_q is non-redundant and, as it follows from Equation (5.16), practically does not depend on the measurand frequency. Moreover, the method of the dependent count has some additional possibilities. It is necessary to give them a quantitative evaluation.

5.7.7 Absolute quantization error Δ_q

Let's consider the time diagrams (Figure 5.12) illustrating one cycle of measurements of unknown frequencies with the periods τ_1, $\tau_2 > \tau_1$ and $\tau_3 > \tau_2$.

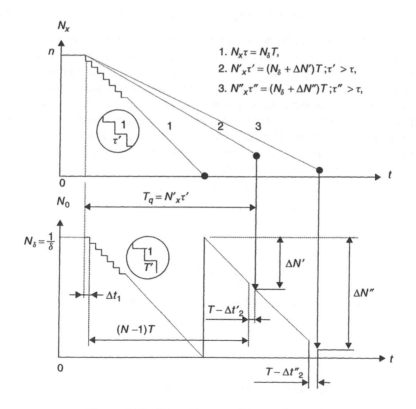

Figure 5.12 Time diagram of conversion cycle

Each next conversion begins at once after the initialization of the counters by the microcontroller. The number $N\delta = 1/\delta$ is written into the counter of the greater frequency $F = 1/T$. Both logical elements are opened and after arrival of the impulse of the lower frequency $f_2 = 1/\tau_2$ the quantization window T_q will be formed. Impulses of the frequency f_2 with the duration $t \ll \tau_{min}$ and impulses of the frequency F after the time $0 \leq \Delta t_1 \leq T$ are calculated in parallel. Therefore the numbers in the counters begin to decrement from 2^{16} in the counter of the lower frequency and from $N\delta = 1/\delta$ in the counter of the greater frequency.

If zeroing happens before the arrival of an impulse of the frequency F_2 (case 2), the impulse count proceeds and the counter decrement begins again from the number N_δ. Through the time interval $(T - \Delta t_2')$ after arrival of the last impulse of the $\Delta N'$ the impulse of the lower frequency comes. It finishes the creation of the quantization window and eliminates the impulse count by both counters. The calculation (Equations (5.18) or (5.19)) begins after reading the numbers Nx' and $\Delta N'$ (the number N_δ is stored in μK's memory). The measurements in the first and the third cases are executed similarly. Thus, with the absolute quantization error $\Delta_q = \Delta t_1 - \Delta t_2$, that does not exceed the $|T|$ for all three cases (Figure 5.12) the following equations are true:

1. $N_x T_x = N_\delta T_0$,

2. $N_x' T_x' = (N_\delta + \Delta N')T_0, \ T_x' > T_x$

3. $N_x'' T_x'' = (N_\delta + \Delta N'')T_0, \ T_x'' > T_x'$ (5.32)

Let's note that the conversion of the unknown frequency f according to the method of the dependent count can be considered as the determination of the average frequency of the sequence consisting of the $(N_\delta + \Delta N)$ impulses with the period T with the help of the rectangular weighting function of the single-level integrating window of Dirichlet. The duration of the window is equal to the integer number of periods τ. The Δt_1 and Δt_2 are independent random components of the quantization error Δ_q, because of non-synchronization of the Dirichlet window with the first and last impulse of the sequence with the period T. These components of errors are distributed according to the uniform, asymmetric distribution law:

$$W(\Delta t_1) = \begin{cases} 0, & \text{if } 0 > \Delta t_1 > T, \\ \dfrac{1}{T}, & \text{if } 0 \leq \Delta t_1 \leq T, \end{cases}$$

$$W(\Delta t_2) = \begin{cases} 0, & \text{if } 0 < \Delta t_2 < -T, \\ \dfrac{1}{T}, & \text{if } 0 \geq \Delta t_2 \geq -T, \end{cases} \qquad (5.33)$$

with the expectation value $|0.5T|$ and the variance $D = T^2/12$.

The absolute error Δ_q is equal to the sum of components Δt_1 and Δt_2. It is distributed according to the symmetrical triangular distribution Simpson law with the following numerical characteristics:

- the expectation value $M(\Delta_q) = M(\Delta t_1) - M(\Delta t_2) = 0$
- the maximum value $\Delta_{q\,max} = \pm T$

- the mean square error

$$\sigma(\Delta_q) = \sqrt{2\int_0^T W(\Delta_q)\Delta_q^2 d(\Delta_q)} = \sqrt{2\int_0^T \frac{1}{T}\left(1 - \frac{\Delta_q}{T}\right)\Delta_q^2 d(\Delta_q)} = \frac{T}{\sqrt{6}}$$

- the variance $D(\Delta_q) = \sigma^2(\Delta_q) = T^2/6$.

5.7.8   Relative quantization error δ_q

The relative quantization error δ_q characterizes more completely the accuracy of the measurement than the absolute one. It is obvious (Figures 5.8, 5.12) that the error arises because the variable time interval formed during the measurement is equal to the quantization window $T_q = N_x \tau = N_x/f$. The interval T_q will be varied at the change of the frequency f. Therefore, the number N of the periods T (Figure 5.12) will also vary within the limits of $0-N_{\max}$ due to the ΔN changes. The latter is determined by the duration of the time interval τ and the frequency F:

$$\Delta N_{\max} = \tau F = \tau/T = N_\delta - 1 \tag{5.34}$$

The error δ_q is distributed according to the symmetrical distribution Simpson law. This error is maximal at the $N = N_{\min} - N_\delta = 1/\delta$ and minimal at the $N = N_{\max} = N_\delta + \Delta N_{\max} = 2N_\delta - 1$. It is determined as:

$$\delta_q = \frac{\Delta_q}{T_q} = \frac{\pm T}{n \cdot \tau} \approx \frac{T}{(N_\delta + \Delta N)\cdot T} = \frac{1}{N_\delta + \Delta N} = \delta\frac{1}{1 + \dfrac{\Delta N}{N_\delta}} = \frac{T}{T_\delta}\cdot\frac{1}{1 + \dfrac{\Delta N}{N_\delta}} \tag{5.35}$$

where $T_\delta = T/\delta$ is the program-specified time of measurement in the mode of the specified speed in addition to the mode with the specified accuracy δ. It is obvious

$$\frac{1}{2N_\delta - 1} = \delta_{q\,\min} \le \delta_q \le \delta_{q\,\max} = \frac{1}{N_\delta} = \delta;$$

$$0.5 \le k = \frac{\delta_q}{\delta} \le 1, \tag{5.36}$$

where k is the coefficient of the change of the error.

The dependencies of the error $\delta_q(\delta, f_x)$ illustrating these features and advantages of the method are shown in Figure 5.13.

Thus, the equation $T_q = n\tau = N_x T_x$ is true for all the frequencies of the measuring range D_1. Each time, when it is defaulted, the specified number of the impulses N_δ is automatically increased on the ΔN due to the extension of the integrating Dirichlet window up to assurance of the multiplicity T_q and $\tau(T_x$ or $T_0)$. Therefore, until $N_\delta \gg \Delta N$ the error δ_q is constant, does not exceed the specified δ and identical at frequencies $f_x < f_0$ and $f_x > f_0$, symmetrical concerning the middle of D_f, in which $f_x = f_0$. The dependence of the error δ_q from the frequency begins to be apparent at increase of the number ΔN. The δ_q decreases on the f_{bound} and F_{bound} boundaries of the range D_f and becomes equal to 0.5δ. The error δ_q continues to decrease in relation to the $1/2N$ at further increase of the number ΔN, if the counter capacity allows.

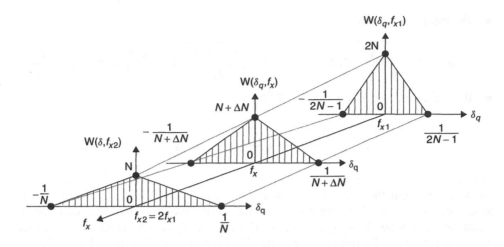

Figure 5.13   Distribution laws of the relative quantization error δ_q

Therefore, due to the method of the dependent count it is possible to realize the self-adopted mode of frequency measurements with the specified relative error δ or the time of measurement T in a broad range of frequencies from infralow up to high, limited by double value of the reference frequency f_0. Thus, the possibility of the extension of the range D_f in the lowest frequencies is determined only by the counter capacity m:

$$\frac{F}{2^{m+1}} \approx \frac{F}{2N_\delta - 1} \le f_{\text{bound}} \le f \le F_{\text{bound}} = F(2N_\delta - 1) \approx 2^{m+1}F \qquad (5.37)$$

In hardware realization the decrease of the lower boundary f_{bound} can be reached due to increase of the used number of counters with the capacity $m = 32, 48, 64$ bits and are more, moreover by decrease of the frequency f_0. For combined and program-oriented realizations, the extension to low and infralow frequencies practically is not limited. It is only determined by the maximum acceptable time $T_q = T_{x\,\text{max}}(N_x = 1)$. The high boundary F_{bound} is determined only by the greatest possible frequency of count for timers/counters and frequency dividers. Thus, the width of the frequency range in the common case is equal to:

$$D_f = \log \frac{F_{\text{bound}}}{f_{\text{bound}}} = \log \frac{f_{\text{max}}}{f_{\text{min}}} = 2^{2(m+1)} \qquad (5.38)$$

5.7.9   Dynamic range

Let's determine the dynamic range of the frequency-to-code converter based on the hardware realization of the method of the dependent count. The decrement counters, used for its realization, limit the width of the range D_f. The high bound F_{bound} is determined by the maximum acceptable frequency f_{clc}, which can be calculated by these counters, and the low f_{bound} by the number of 16-bit counters. Therefore

$$\Delta N_{\text{max}} = 2^{16n_c} \qquad (5.39)$$

Table 5.1 Dynamic range at $\Delta\delta_q = 0.5\delta$

```
              2f_bound                        2F_bound                           D_f
  δ    10^-7  10^-6  10^-5  10^-4   10^-7  10^-6  10^-5  10^-4    10^-7  10^-6  10^-5  10^-4
f_0(Hz)

10^6   10^-1---------D_f---------10^13                          10^14
           10^0---------D_f---------10^12                              10^12
               10^1---------D_f---------10^11                                10^10
                   10^2---------D_f---------10^10                                  10^8

10^7   10^0---------D_f---------10^14                           10^14
           10^1---------D_f---------10^13                              10^12
               10^2---------D_f---------10^12                                10^10
                   10^3---------D_f---------10^11                                  10^8

10^8   10^-1---------D_f---------10^15                          10^14
           10^2---------D_f---------10^14                              10^12
               10^3---------D_f---------10^13                                10^10
                   10^4---------D_f---------10^12                                  10^8
```

With regard for the above, the decrease of the error δ_q relative to the specified δ does not exceed the values:

$$\frac{\delta_q}{\delta} = \frac{1}{1+\dfrac{\Delta N}{N_\delta}} = \begin{cases} 1, & \text{if } \Delta N = 0, \\ 0.5, & \text{if } \Delta N = \Delta N_{max}, \\ 0.5 < k < 1, & \text{if } 0 < \Delta N < \Delta N_{max} \end{cases} \qquad (5.40)$$

Having accepted for simplification $\Delta N_{max} \approx \tau F$, $N_x = 1$, and $N \approx 2N_\delta$, we receive the formulas for determination f_{bound}, F_{bound} and D_f:

$$f_{bound} = 0.5 \cdot f_0 \cdot \delta; \quad F_{bound} = 0.5 \cdot f_0/\delta; \quad D_f = 1/\delta^2 \qquad (5.41)$$

The calculated results according to the Equation (5.41) for that case, when $\Delta\delta_q = 0.5\,\delta$, are adduced in Table 5.1. The width of the frequency range D_f is shown in Table 5.2.

The possibilities of the symmetrical extension and shifting of the range D_f by the decrease of the error δ and the change of the frequency f_0 are obvious from these tables. By the decrease of the error δ the range D_f symmetrically extends concerning the middle—the frequency f_0, for which:

$$f_x = f_0 = 0.5 \cdot (F_{bound} - f_{bound}),$$

$$\delta_q = \delta_{q\,max} \leq \delta \qquad (5.42)$$

The time T_q is minimum possible:

$$T_q = N_x \cdot T_x \approx N_\delta \cdot T_0 = \frac{1}{\delta \cdot f_0} = T_{q\,min} \qquad (5.43)$$

Table 5.2 The width of the frequency range D_f

δ	10^{-2}	10^{-3}	10^{-4}	10^{-5}	10^{-6}	10^{-7}	10^{-8}	10^{-9}
D_f	10^4	10^6	10^8	10^{10}	10^{12}	10^{14}	10^{16}	10^{18}

In frequency counters based on traditional methods of measurement, the relative quantization error δ_q increases according to the hyperbolic law by the decrease of the measurand frequency f_x or the period T_x:

$$\delta_q = \delta_{\text{frequency}} \frac{f_{x\text{Nom}}}{f_x},$$

$$\delta_q = \delta_{\text{period}} \frac{T_{x\text{Nom}}}{T_x} \tag{5.44}$$

Therefore, their range D_f is much narrower. It is asymmetric concerning the frequency f_0 and equals only to one decade. Thus $\delta_q = 10\delta_{\text{frequency}}$ and $\delta_q = 10\delta_{\text{period}}$ at $f_x = 0.1 f_{x\text{Nom}}$ $T_x = 0.1 T_{x\text{Nom}}$. Unlike the method of the dependent count the quantization window T_q and the quantization frequency f_q remain constant within the limits of one decade:

$$T_q = \frac{1}{f_{x\text{Nom}}\delta} = \text{const},$$

$$f_q = f_0 = \frac{1}{T_{x\text{Nom}}\delta} = \text{const} \tag{5.45}$$

They are redundant for all measurand frequencies and periods from the range D_f, except the nominal one. Further increase of the range D_f is possible. However, it requires additional hardware and special technical solutions.

The method of the dependent count allows to set not only the maximum error δ, that does not depend on the measurand frequency, but also the coefficient $k > 0.5$ (for example, $k = 0.9$–0.98 or lower). This coefficient can be used for limitation from below the error δ_q (in the boundaries $k\delta < \delta_q < \delta$) and from above the time T_q. However, the range $D_f(k\delta)$ is narrowed down. In this case, Equations (5.41) should be rewritten:

$$f_{x\,\min}(k\delta) = \frac{k}{1-k} f_0\delta > f_{\text{bound}},$$

$$f_{x\,\max}(k\delta) = \frac{1-k}{k\delta} f_0 < F_{\text{bound}} \tag{5.46}$$

$$D_f(k\delta) = \left(\frac{1-k}{k\delta}\right)^2 < D_f \tag{5.47}$$

The ways of the extension of the range $D_f(k\delta)$ or its shifting remain similar to those explained above. The δ decreasing does not require additional hardware.

5.7.10 Accuracy of frequency-to-code converters based on MDC

The accuracy of conversion and metrological reliability are the most important performances of a smart sensor. The main components of static and dynamic errors of the method of the dependent count are shown in Figure 5.14.

The static error is stipulated by the trigger error, the reference frequency instability, the quantization error and the calculating error. The latter is determined by calculation

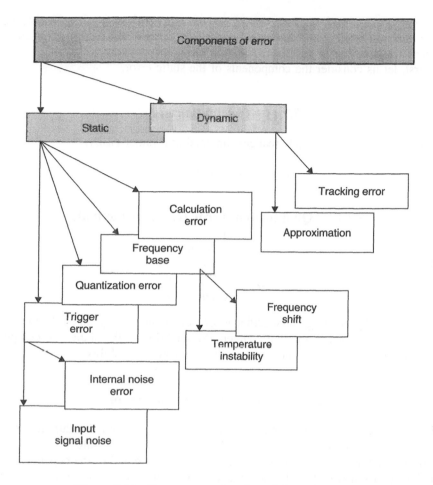

Figure 5.14   Components of static and dynamic errors

of the result of measurement according to the specified formulas with the use of the numbers N_x, ΔN and consequent rounding of the result.

The dynamic error is stipulated by the change of the measurand frequency during the time interval T_q and between the cycles of measurement. The method of the dependent count organizes the two-channel mode of continuous frequency measurement without loss of speed and accuracy. In this case, the dynamic error will be determined only by duration of the quantization window T_q.

The maximal $\delta_{f\,\text{max}}$ and the mean-square static errors σ_f are determined as:

$$\delta_{f\,\text{max}} = \frac{1}{N_x}\delta_{\text{TriggerErrormax}} + \delta_{q\,\text{max}} + \delta_{0\,\text{max}} + \delta_{\text{Calculationmax}} \tag{5.48}$$

$$\sigma_f = \sqrt{\frac{\sigma_{\text{TriggerError}}^2}{N_x^2} + \sigma_q^2 + \sigma_0^2 + \sigma_{\text{Calculation}}^2}, \tag{5.49}$$

where $\delta_{TriggerErrormax}$ and $\sigma_{TriggerError}$, $\delta_{q\,max}$ and σ_q, $\delta_{0\,max}$ and σ_0, $\delta_{Calculation}$ and $\sigma_{Calculation}$ are the maximal relative and the mean-square trigger error, the quantization error, the reference frequency error and the calculation error, respectively.

At first, let us consider the components of the static error.

5.7.11 Calculation error

The use of modern μP and μK realizes the condition rather easily:

$$\delta_{Calculationmax} \ll \delta_{min} \tag{5.50}$$

5.7.12 Quantization error (error of method)

$$\delta_{q\,max} = \frac{1}{N_\delta} = \delta,$$

$$\sigma_q = \frac{1}{N_\delta\sqrt{6}} = \frac{\delta}{\sqrt{6}} \tag{5.51}$$

It is determined by the program-specified error δ and does not depend on the measurand frequency. It is supposed, that the counters do not bring in the additional component to this error because of the time and temperature trigger instability.

5.7.13 Reference frequency error

The reference clock quartz generator (with the period T_0 and the duration $\tau \ll T_0$ (Figure 5.12) is a source of the error δ_0. Its influence on the result follows from the definition of the measurand frequency according to the method of the dependent count (Equations (5.18) and (5.19)). It is stipulated by the long-term and short-term frequency instability:

$$f_0 = f_{0Nom}(1 + \delta f_0) \tag{5.52}$$

The change of the frequency f_0 at the constant measurand frequency f_x will result in changing of the number N, and consequently, of the result. In the mode, when $f_x > 0$, the time interval T_q will be varied. Thereof the result of measurement will be also changed.

The long-term changes depend on the systematic deviations of the frequency f_0 stipulated by the time drift. It is $10^{-9}-10^{-11}$ for one day, and causes the systematic component of the error δ_0.

The short-term changes depend on the frequency fluctuation. They cause the random component of the error δ_0 because of short-term temperature and time drifts.

The systematic component of the error δ_0 can be reduced by the correction of the frequency error up to the random level. In practice it is necessary to take into account only temperature and time instabilities, that is equal to $\approx 10^{-7}-10^{-9}$ and is determined by the quality factor of a quartz resonator. The last one can achieve the value 50×10^{-6} and higher for precision quartz resonators. Due to this the random variation of the frequency of the reference generator can be not worse than 10^{-11}.

Therefore, the error δ_0 should be considered as a random error with the uniform distribution law within the boundaries $\pm\delta_{0\,max}$:

$$\delta_{0\,max} = 10^{-7}\text{--}10^{-9}$$

$$\sigma_0 = \frac{\delta_{0\,max}}{\sqrt{3}} \tag{5.53}$$

The temperature control and temperature compensator for quartz generators reduce the error $\delta_{0\,max}$ and increase considerably the frequency stability.

The use of external generators with more stable reference frequency, connected with the atomic frequency standard is rather perspective. The caesium and rubidium frequency standards are widely used in this case. Their instability does not exceed $(2\text{--}3) \times 10^{-12}$ and $(1\text{--}3) \times 10^{-11}$ accordingly. For example, the use of the device OFS2/2 from Gould Advance Corporation guarantees the frequency stability not worse than $\pm 1 \times 10^{-10}$ during 1000 seconds. It determines the maximal value of the period $T_x = T_q$, which can be measured with such stability.

As the relative error δ can be varied over a wide range, each time, when $\delta > \delta_{0\,max}$ in 3–5 times it is possible to consider the latter very little and not take into account Equations (5.48) and (5.49).

5.7.14 Trigger error

The input blocks F1 and F2 (Figure 5.11), consist of an attenuator, a controlled amplifier and a Schmitt trigger. They select the period of the input signal and form the sequence of rectangular impulses with the period T'_x and the duration $t \ll T'_{x\,min}$, and ensure the triangular unbiased distribution law of the quantization error δ_q. Because of incomplete signal processing limited by possibilities of the F1 and F2, the period T'_x differs from the real period T_x of the frequency f_x on the value $\pm\Delta T_x = \pm\delta_{TriggerError}\,T_x$. The quantization window $T'_q = N'_x T'_x$ will differ from the conventional true value T_q on the value $\pm N'_x \Delta T'_x$ which can be large. Therefore, the accuracy of the frequency-to-code converter can be determined by the error $\delta_{TriggerErrormax}$, instead of the errors $\delta_{0\,max}$ and $\delta_{q\,max}$. Especially it is characteristic for conversion of low frequencies of the harmonic signal ($N_x \to 1$).

The first of two components of the $\delta_{TriggerError}$ is the trigger level setting error due to deviation of the actual trigger level from the set trigger level. The second component of the error $\delta_{TriggerErrormax}$ occurs when the measurement starts or stops too early or too late because of noise on the input signal. At rather low noise level the influence on the selection of the period T_x can be reduced and even eliminated by a choice of the trigger hysteresis. The smooth narrow-band noise changes the moment of operation of the input device and can also reduce to greater difference of the duration T'_x from the T_x. In all cases, the noise appears both at the beginning and at the end of each period of the input signal. By the broadband noise, it is difficult to determine the period T_x, as the time intervals between adjacent operations of the input device do not characterize the period T_x any longer. Special measures, for example, the usage of input filters with the cut-off frequency are necessary for the elimination of this influence:

$$f_c \approx \frac{\sqrt{V_m^2 - V_q^2}}{T_x \cdot V_{Neff}} \tag{5.54}$$

Thus, fluctuations of the operation level and noise reduce work of the trigger up to or after the necessary moment. Therefore because of such a false operation of the Schmitt trigger, the time interval $T_q' = T_q \pm N_x' \Delta T x'$ will be quantized by impulses with the period T_0 if $f_x \le f_0$ or $T_q = N_x T_0$ by impulses with the period $T_x' = T_x \pm \Delta T_x$, if $f_x > f_0$ instead of the interval $T_q = N_x T_x$. If special measures are not employed, there will be an absolute error of measurement of the frequency $\Delta f_x'$.

The mean-square value of the deviation ΔT_x of the period T_x of a harmonic signal and the relative error of definition of its conventional true value are determined as in [95]:

$$\sigma = \pm \frac{0.225 \cdot V_{\text{Neff}}}{\sqrt{V_m^2 - V_q^2}} \cdot T_x,$$

$$\delta_{\text{TriggerError}} = \pm \frac{22.5 \cdot V_{\text{Neff}}}{\sqrt{V_m^2 - V_q^2}}\%, \tag{5.55}$$

where V_{Neff} is the effective noise voltage; V_m is the amplitude of the sinusoidal voltage of the measurand frequency.

Even the low noise reduces in significant errors $\delta_{\text{TriggerError}}$. So, for example, if $V_m = 10$ V, $V_{\text{Neff}} = 0.1$ V (1% in relation to the input signal) and the quantization level $V_q = 0.3$ V, the error will be $\delta_{\text{TriggerError}} \approx \pm 0.225\%$. As it was determined at a signal/noise ratio equal to 40 dB the error will be $\delta_{\text{TriggerError}} = 0.3\%$.

With the aim of decreasing the $\delta_{\text{TriggerError}}$ it is necessary to quantize some periods T_x'. In this case, the error $\delta_{\text{TriggerError}}$ will be decreased N_x times. In the case when higher frequencies are measured and $N_x \gg 1$, the error $\delta_{\text{TriggerError}}$ can be neglected.

By the measurement of the frequency of impulse sequence, the error $\delta_{\text{TriggerError}}$ will decrease. In this case, it depends on the number N_x, the period T_x, the set trigger level, its drift Δ and the wavefront steepness S [95]:

$$\delta_{\text{TriggerError}} \approx \pm \frac{100 \cdot \Delta}{N_x S \cdot T_x}\% \tag{5.56}$$

The error $\delta_{\text{TriggerError}}$ can also be neglected by measuring frequency f_x of pulses with steep wavefronts.

Now let us consider the dynamic error and its components. By frequency measurements, this error essentially depends on the dynamic properties of the researched process and the speed of the method of measurement. The decrease of the dynamic error is possible by the increase of speed and continuous correction of the measurement results.

The method of the dependent count has the highest speed at measurement of all frequencies of the range D_f and consequently allows reduction of the dynamic error for the same frequencies of researched physical values, as well as for traditional methods of measurement. The main components of the dynamic error are the tracking error and the approximating error.

The first component depends on the time T_q. The result of the measurement of duration of the period or several periods of low frequencies composing the time T_q, corresponds to their average value on the time interval equal to their duration. The inevitable availability of such an average and also the change of the researched value during the latency time if it is not eliminated, is the reason of the rise of the so-called average error.

The second component is the approximation error of the continuously varying value of the lattice function with the appropriate approximation between points of its discrete values. The approximation error is determined by the digitization interval and a kind of approximation. In one-channel and multichannel measuring instruments the digitization interval cannot be lower than the time of the measurement T_q.

The decrease of the dynamic error can be reached by minimization of the time T_q by the choice of the algorithms, the eliminating latency time of the next cycle of conversion.

The dynamic error should not exceed the static. Under this condition, the acceptable time T_q at the known frequency of change of the researched value and its maximum value at the selected T_q are determined. This algorithm can also be used at higher frequencies of the range D_f, if the time of the measurement is large, even in the case of the method of the dependent count.

5.7.15 Simulation results

The quantization error δ_q is the function of many variables and is determined by the given relative error δ and the frequencies f and F. This function $\delta_q(\delta, F, f)$ is piecewise-smooth and has a final number of discontinuities of the first kind. In this analysis of the family of characteristics we will start from the function $\delta_q(\delta, F)$ at $f = $ const. All characteristics of the given family are symmetric concerning the middle of the range $D_f(\delta_i, F)$. In this point $\Delta N = 0$ and $f = F$. Therefore in these points and their neighbourhoods $\delta_q = \delta_{q\,max} = \delta$. Change of the ordinate δ_i for any of these functions at $F = $ const reduces in its parallel moving up (down) at increase (decrease) of the error δ_i. By this, the range D_f, within the limits of which the condition (5.36) is satisfied varies. If the counter capacity is sufficient, the frequency measurement outside the range D_f with the errors $\delta_q < 0.5\ \delta$ is possible. Simultaneously the number N_{break} of discontinuity of the first kind will be varied. By this, their number is increased with decrease of the error δ.

The change of the frequency F at $\delta = $ const reduces the shift of the characteristic of the family towards low frequencies, at its decrease, and towards high ones, at its increase, without the change of the range D_f. However, the maximum measurand frequency $f_{max} \le 2F$ therefore will be varied.

Now let's consider the characteristic $\delta_q(f, F)$ at $\delta = $ const. If the frequency f varies relatively to the F, the dependence of the error δ_q from the frequency f becomes apparent in a various degree from some value. Due to the self-adopting of the method this dependence is expressed much more poorly than in the traditional methods of comparison. The number ΔN is increased from 1 up to ΔN_{max} symmetric relatively to the frequency $F = f_0$ boundaries f_{bound} and F_{bound} of the range D_f because of the automatic change of the Dirichlet window. The error δ_q is decreased from δ up to the minimal value $\delta_{q\,min} = 0.5\delta$. In the common case, on the boundaries D_f:

$$\Delta N_{max} = N_\delta - 1 = \begin{cases} T_x f_0 - 1, & \text{if } f_x \le f_0, \\ T_0 f_x - 1, & \text{if } f_x > f_0 \end{cases} \tag{5.57}$$

Let us investigate the required frequency dependence of the error δ_q, its jumps in the points N_{break}, the boundaries of the discrete steps and their value $\Delta\delta_q$ in more detail. For simplification we shall consider only the first half of the D_1 of the range D_f of the function $\delta_q(f, F)$, in which $f = f_x \leq F_0 = F$ and $\delta = \text{const}$ and $f_0 = \text{const}$. Due to the symmetry, the obtained results will also be true for the second half of the D_2 of the range D_f.

The mathematical analysis of the function $\delta_q(f_x, f_0)$ on continuity has confirmed the availability of discontinuities of the first kind and has shown, that by the change of the frequency f_x in the last interval $(N_x - 1)T_x \ldots N_x T_x$ in the quantization window T_q the change of the number N is a reason for jumps in the points N_{break}. It makes the following conclusions.

(1) The values of the jumps $\Delta\delta_{qi}$, the number $\Delta N_i = t_i' f_0$ and the time intervals $0 \leq t' \leq T_x$ are decreased and tend to zero by the increase of the frequency f_x. Therefore in this greater part of both halves of the range D_f contiguous to its middle, the plot of the function $\delta_q(f_x, f_0)$ presents a significant number of slanting sites of hyperbola of various lengths, divided by the jumps $\Delta\delta_{qi}$ in the points N_{break} with co-ordinates:

$$\delta_{qi} = \delta \frac{N_\delta}{N_\delta + \Delta N_i} = \delta \frac{N_\delta}{N_\delta + t_i' \cdot f_0} \tag{5.58}$$

In accordance with the decrease of the number ΔN_i the value of the jumps $\Delta\delta_{qi}$ decreases, and the segments of the hyperbolas $\delta_q(f_x, f_0)$ juxtapose up to full junctions and coincide with the plot of a hypothetical (ideal) method. The latter represents a direct line, parallel to the axis of frequencies with the ordinate δ. Therefore the error δ_q practically does not depend on the frequency f_x (Figure 5.15(a). Such a dependence is saved by the change of the frequency f_0 over the wide range and the error $\delta = 0.0005 - 0.00005$.

(2) By the decrease of the frequency f_x the Dirichlet window is extended. The duration t_i', the number ΔN_i and $\Delta\delta_{qi}$ between adjacent partial segments of hyperbolas composing the function $\delta_q(f_x, f_0)$ is increased. On the low bound f_{bound} the number $\Delta N = \Delta N_{\text{max}} = N_\delta - 1$, the error $\delta_q = \delta_{q\min} = 0.5\delta$, and $\Delta\delta_q = 0.5\delta$.

The simulation results of the functions $\delta_q(f_x, f_0)$ for a wide range of low frequencies are shown in Figure 5.15(b). The segment of the piecewise-smooth function composed from the partial segments of the hyperbolas at $f_0 = 20$ kHz and $f_x \cong 1$–8 Hz is shown in the foreground of the figure. The analysis of these functions for other values of f_0 has confirmed the dependence of the error δ_q from the f_x in the extreme minority of the range D_1 that exists and weakens by the increase of the f_0.

Thus, in both cases the error δ_{qi} never exceeds the given value δ_i in the whole range D_f of measurand frequencies from f_{bound} up to F_{bound}. Each function from the set $\delta_{qi}(f_x, f_0)$ at $\delta_i = \text{const}$ is piecewise-smooth. It consists of the final number of partial segments of hyperbola divided by jumps of the various value $\delta\Delta_{qi}$ in the discontinuities N_{break} of the kind. Their number considerably increases by the decrease of the program-specified error δ.

Let's determine the boundaries of discontinuities of the piecewise-smooth function $\delta_q(\delta, f_x)$ and the value of the jumps $\Delta\delta_q$ in the points N_{break} between them. For that purpose, we shall take the advantage of a known technique and Equation (5.58). By the increase of the frequency f_x and approximation to the point N_{break} from the left

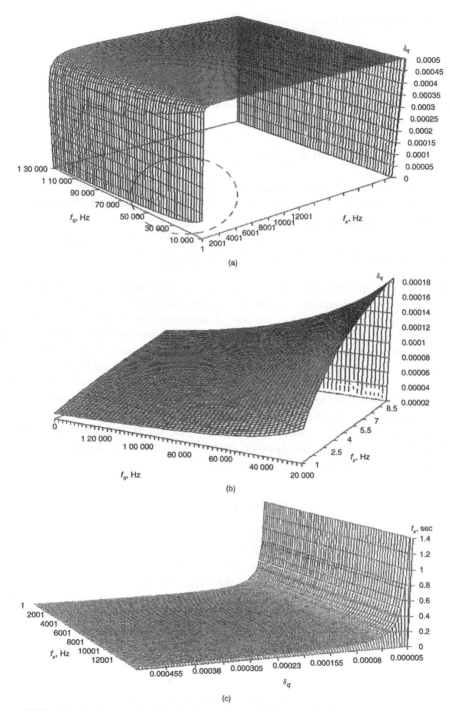

Figure 5.15   Simulation results: (a) the dependence of the quantization error on the frequency in the whole range of measurand frequencies; (b) the dependence of the quantization error on frequency in the low frequency range; (c) the dependence of time of measurement on frequency and error of measurement

the error δ_q goes to the limit:

$$\delta_q (N_{\text{break}} - 0) = \lim_{N \to N_\delta} \delta \frac{N_\delta}{N} = \delta \qquad (5.59)$$

and the left-side boundary of the function will be formed.

By the decrease of the frequency f_x and approximation to the point N_{break} from the right the error δ_q goes to the limit:

$$\delta_q (N_{\text{break}} + 0) = \lim_{(N-1) \to N} \delta \frac{N_\delta}{N - 1} = \delta \frac{N_\delta}{N} = \delta \frac{N_\delta}{N_\delta + \Delta N} \qquad (5.60)$$

and the right-side boundary of the function will be formed.

As $\delta_q (N_{\text{break}} - 0) \neq \delta_q (N_{\text{break}} + 0)$, the function has the discontinuity in the point N_{break} and the finite jump $\Delta \delta_q$, determined as:

$$\Delta \delta_q = \delta_q (N_{\text{break}} + 0) - \delta_q (N_{\text{break}} - 0) = \delta \frac{N_\delta}{N_\delta + \Delta N} - \delta = -\delta \frac{\Delta N}{N_\delta + \Delta N} \qquad (5.61)$$

The negative sign specifies that the left boundary is always more than the right one and by the decrease of the frequency f_x the error δ_q decreases and the jump $\Delta \delta_q$ increases. By this, the time T_q is increased at the cost of the extension of the Dirichlet window. As follows from Equations (5.17) and (5.18), the Dirichlet window practically does not depend on the measurand frequency. It is variable and non-redundant for each of the frequencies from the range D_1 and is set up automatically during the measurement. Therefore, in comparison to all known methods, the method of the dependent count is a self-adopted method. Due to this, the method has high metrological performances. The real function $\delta_q (f_x, f_0)$ coincides in greater part of the range with ideal, hypothetical. The simulation results of the set of the functions $t_x (\delta, f_x)$ at $f_0 = \text{const}$ are shown in Figure 5.15. They confirm high dynamic properties of the method, due to which the dynamic error of the measurement is minimal possible.

5.7.16   Examples

In order to compare the method of the dependent count with the standard counting methods as well as with other advanced methods for the frequency-to-code conversion let us determine the coefficient of variation for the quantization error $\alpha = \delta_{q\,\text{max}} / \delta_{q\,\text{min}}$ for these methods. So, for the method of the dependent count if f_x is lower of two frequencies ($f_x = f$), and f_0 is greater of the frequencies ($f_0 = F$), i.e. $f_x < f_0$, from the following equations

$$\delta_{\text{max}} = \frac{1}{N_{\text{min}}} = \frac{1}{N_\delta}$$

$$\delta_{\text{min}} = \frac{1}{N_{\text{max}}} = \frac{1}{N_\delta + \Delta N_{\text{max}}}, \qquad (5.62)$$

and taking into account Equation (5.34) we will have

$$\alpha = \frac{N_\delta + \Delta N_{\text{max}}}{N_\delta} = 1 + \frac{1}{N_\delta} \cdot \frac{F}{f} \qquad (5.63)$$

or

$$\alpha_1 = 1 + \frac{1}{N_\delta} \cdot \frac{f_0}{f_x} \tag{5.64}$$

At $f_x > f_0$

$$\alpha_2 = 1 + \frac{1}{N_\delta} \cdot \frac{f_x}{f_0} \tag{5.65}$$

Let's determine how many times the quantization error will be varied by the measuring frequency $f_x = 2$ Hz, if $f_0 = 10^6$ Hz and $N_\delta = 10^6$ ($\delta = 10^{-6} \times 100\% = 0.0001\%$). From Equation (5.64) we have

$$\alpha_1 = 1 + \frac{1}{10^6} \cdot \frac{10^6}{2} = 1.5,$$

i.e. the greatest error $\delta_{max} = 1/N_\delta = 10^{-6}$, and the lowest $\delta_{min} = 0.67 \times 10^{-6}$. As the greatest quantization error for the method of the dependent count is constant for any measurand, it is possible to characterize the possible range of variation of this error in the specified measuring range of frequencies by the coefficient of variation α. From this example it follows that the error variation is not more than 1.5 times in the frequency range $2–10^6$ Hz (the time of the measurement is constant for the given quantization error). By the use of the standard direct counting method or the indirect method measuring period, the variation of the quantization error will be 500 000 (at the same time of measurement).

Let us consider another example. It is necessary to measure $f_x = 10^4$ Hz at $f_0 = 10^6$ Hz and $N_\delta = 10^4$. According to the method of the dependent count the maximum time of conversion (5.17) is

$$T_{q\,max} = \frac{10\,000 + \dfrac{1}{10^4}}{10^6} \approx 0.01 \text{ s.}$$

In turn, according to the standard counting method, the time of measurements necessary for the same accuracy is $T_{q\,max} = 0.5$ sec. In other words, the time of measurement for the proposed method is non-redundant in all specified measuring ranges of frequencies. In the standard counting method, the time of measurement is redundant, except for the nominal frequency. Moreover, for the method of the dependent count the time of measurement can be varied during measurements depending on the assigned error.

5.8   Method with Non-Redundant Reference Frequency

Power consumption and dissipation on the elements becomes actual by smart sensor creation. The dynamic average power of a circuit can be given as

$$P_{avr} = C_{eff} \cdot V_{DD}^2 \cdot f_{clc}, \tag{5.66}$$

where V_{DD} is the supply voltage; f_{clc} is the clock frequency; C_{eff} is the effective capacitance of the circuit [126]. V_{DD} and C_{eff} are constant for the specific integrated circuit and technology. For many smart sensors V_{DD} is reduced up to 2.8–3.5 V. The power consumption is directly proportional to the system clock. If the clock speed doubles, the current doubles. Obviously, power can be saved by operating the device at the lowest possible clock speed [127]. The reduction of f_{clc} is in contradiction to that necessary to increase the metrological performances for precise measurements. In the majority of methods for frequency-to-code conversion, the reference frequency f_0 directly influences the conversion accuracy: the reduction of f_0 will increase the quantization error. In order to eliminate this inconsistency, another self-adapting method of frequency-to-code conversion with non-redundant reference frequency f_0 was proposed [114].

The essence of this method consists of the following. The embedded microcontroller or the arithmetic unit calculates the value of the necessary reference frequency according to the given quantization error δ_i:

$$f_{0i} = \frac{k}{\delta_i},$$ (5.67)

where $k = 1/T_0 = \mathrm{const}$ (T_0 is the first reference gate period). The reference frequency f_{0i}, which is received by division/multiplication of the clock frequency f_{clc} and the measurand frequency of a quartz generator is calculated by two counters. The time of measurement is equal to the integer number of periods of f_x. This frequency is calculated similarly, as for the majority of advanced methods for frequency-to-code conversion:

$$f_x = \frac{N_1}{N_2} \cdot f_{0i}$$ (5.68)

The quantization error does not exceed that given beforehand:

$$\delta_i = \frac{1}{T_0 \cdot f_{0i}}$$ (5.69)

Similar to the other advanced methods for frequency-to-code conversion, this method ensures a constancy of the quantization error in the total specified measuring range of frequencies–from infralow up to high frequencies. Besides that, the reference frequency f_{0i} is non-redundant and determined by the given quantization error δ_i. Thus, by the use of the method with the non-redundant reference frequency in smart sensors it is possible to receive further benefits. It has a high accuracy of measurement, a constancy of the quantization error and a reduction of the power dissipation during the conversion by the use of the lower (on the average by 1–2 order) reference frequency.

Modern achievements in microelectronics easily realize this method with the minimum possible hardware. The MSP430 microcontroller family for metering applications from Texas Instruments [128] is the most appropriate for realization of such a method for frequency-to-code conversion. These microcontrollers have capabilities that make them suitable for low-power embedded applications. As follows from this reference, one of the main characteristics of a power-efficient microcontroller core is the flexible clocking. This performance makes the MSP430 microcontroller family ideally appropriate.

The system clock frequency f_{system} in the MSP430 microcontroller family depends on two values:

$$f_{system} = N \cdot f_{crystal},$$ (5.70)

where $N(3 - 127)$ is the multiplication factor; $f_{crystal}$ is the frequency of the crystal (normally 32 768 Hz). The normal way to change the system clock frequency is

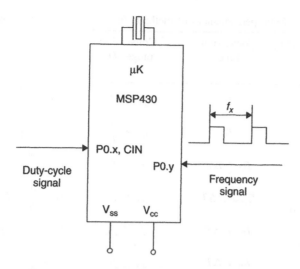

Figure 5.16   Connection of frequency-time domain signal to the MSP430

to change the multiplication factor N. The system clock frequency control register SCFQCTL is loaded with $(N - 1)$ to achieve the new frequency [129].

Figure 5.16 shows the connection of the frequency signal to the MSP430 microcontroller. It can be connected to any of the eight inputs of Port0 and counted via the interrupt. If the frequencies to be measured are above 30 kHz then the universal timer/port or the 8-bit interval timer/counter may also be used for counting. The first reference time window is formed by the basic timer.

If the information to be converted is represented by pulse distances or pulse widths then it is also easy to be converted with the MSP430. The left part of Figure 5.16 shows how to do this.

The signal to be converted is connected to one of eight inputs of Port0. Each one of these I/Os allows interrupt on the trailing and on the leading edge. With the basic timer an appropriate timing is selected for the required resolution and the conversion is made. The universal timer/port may be used for this purpose too: the pulse to be measured is connected to the pin CIN and the time measured from edge to edge.

Even better resolution is possible with the Timer_A. The input signal is connected to one of the TA-inputs and the capture register is used for the time measurements.

5.9   Comparison of Methods

For choice of method for frequency-to-code conversion, it is expedient to prefer one that has high metrological characteristics and simple realization by its universality. The main performances of all the methods considered above are adduced in Table 5.3. Here number 1 is the indirect counting method (period measurement); 2 is the standard direct counting method (frequency measurement); 3 is the ratiometric counting method; 4 is the reciprocal counting method; 5 is the M/T counting method; 6 is the constant elapsed time (CET) method; 7 is the single-buffered and double-buffered method; 8 is

Table 5.3 Main performances of methods for frequency-to-code conversion

Method	Quantization error, δ_q	Conversion time	Conversion range, D_f	Calculation of result	Quantization error variation, α
1.	$\dfrac{f_x}{f_0}$	T_x	$f_x \ll f_0$	$f_x = \dfrac{f_0}{N_2}$	$f_{x\,max}/f_{x\,min}$
2.	$\dfrac{1}{T_0 \cdot f_x}$	T_0	$f_x \gg \dfrac{1}{T_0}$	$f_x = \dfrac{N_1}{T_0}$	$f_{x\,max}/f_{x\,min}$
3.	$\dfrac{1}{f_0 \cdot T_{01}}$	$T_{01} + \Delta T$	$f_x \leq f_0$	$f_x = \dfrac{N_1}{N_2} \cdot f_0$	1–1.5
4.	$\dfrac{1}{f_0 \cdot T_{count}}$	$T_{count} + \Delta T$	$f_x \leq f_0$	$f_x = \dfrac{N_1}{N_2} \cdot f_0$	1–1.5
5.	$\dfrac{1}{f_0 \cdot T_{01}}$	$T_{01} + \Delta T$	$f_x \leq f_0$	$f_x = \dfrac{N_1}{N_2} \cdot f_0$	1–1.5
6.	$\dfrac{1}{f_0 \cdot T_{01}}$	$T_{01} + \Delta T$	$f_x \leq f_0$	$f_x = \dfrac{N_1}{N_2} \cdot f_0$	1–1.5
7.	$\dfrac{1}{f_0 \cdot T_0}$	T_0	$f_x < f_0$	$f_x = \dfrac{N_1}{N_2} \cdot f_0$	1–2
8.	$\dfrac{1}{f_0 \cdot T_0}$	T_0	$f_x < f_0$	$f_x = \dfrac{N_1}{N_2} \cdot f_0$	1–10
9.	$\dfrac{1}{T_{01} \cdot f_{0i}}$	$T_{01} + \Delta T$	$f_x \leq f_{0i}$	$f_x = \dfrac{N_1}{N_2} \cdot f_{0i}$	1–1.5
10.	$\dfrac{1}{N_\delta}$	$\dfrac{N_\delta + \Delta N}{f_0} = \dfrac{N_2}{f_0}$	$f_{x\,min} < f_0 \leq f_{x\,max}$	$f_x = \dfrac{N_1}{N_2} \cdot f_0$	1–1.5

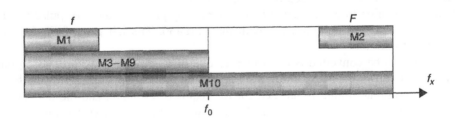

Figure 5.17 Width frequency ranges D_f for different methods of frequency-to-code conversion

the DMA transfer method; 9 is the method with the non-redundant reference frequency; 10 is the method of the dependent count.

As it can be seen from the table, the majority of modern advanced counting methods overcome the demerit inherent to classical conversion methods. Namely, it means the inconstancy of the quantization error in all specified ranges of converted frequencies. However, most of them (methods with the constant conversion time as well as with the slightly varied conversion time) have a redundant conversion time for all frequencies, except the nominal one. The exceptions are the method of the dependent count and the classical indirect counting method (period measurement). Only one of methods, namely the method of the dependent count measures the frequency $f_x \geq f_0$, and only one method, namely method 9 has the non-redundant reference frequency.

Conversion ranges for all considered methods are shown in Figure 5.17.

5.10   Advanced Method for Phase-Shift-to-Code Conversion

The phase shift φ_x between two periodic sequences of pulses with the period T_x can be converted by the *method of coincidence* [104]. Time diagrams of the method are shown in Figure 5.18. In this case the number N_1 pulses with the period T_0 and the number N'_1 pulses of the first sequence T_x between coincident pulses of these sequences is counted. Then

$$N_1 \cdot T_0 = N'_1 \cdot T_x \tag{5.71}$$

Similarly, the number N_2 of pulses T_0 and the number N'_2 of the second sequence with the period T_x, shifted on the t_x and taking place between the first moment of coincidence of the first pair of pulses and the nearest moment of coincidence of the second pair of the pulse are counted. Then

$$N_2 \cdot T_0 = N'_2 \cdot T_x + t_x. \tag{5.72}$$

From these two equations, we receive the formula for the phase-shift calculation

$$\varphi_x = \left(\frac{N'_1 \cdot N_2 - N'_2 \cdot N_1}{N_1} \right) \cdot 360, \tag{5.73}$$

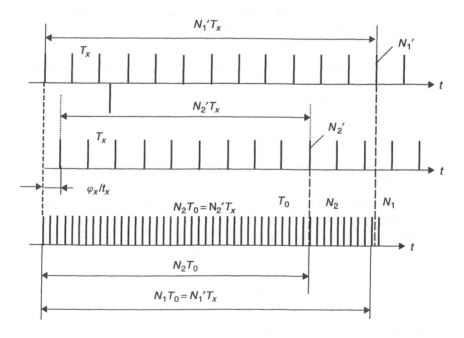

Figure 5.18   Time diagrams of the method of coincidence for phase-shift-to-code conversion

and for the converted time interval

$$t_x = \left(\frac{N_1' \cdot N_2 - N_2' \cdot N_1}{N_1} \right) \cdot T_0. \tag{5.74}$$

The analysis of Equations (5.71) and (5.74) shows that φ_x and t_x do not depend on the period T_x. Conversion errors will be determined mainly by pulses duration only. For reduction of these errors, the method of forming of pulse packets of coincidences can be used. Thus, the absolute error of measurement for t_x can be reduced up to 0.5×10^{-12} s and the absolute conversion error for the phase shift φ_x up to $0.05°$ at 1 MHz.

Summary

Due to advanced methods for frequency-to-code conversion, it is possible to achieve a constant quantization error for all ranges of converted frequencies.

The method of the dependent count is optimal for microcontroller-based frequency-to-code converters of a new generation for the measurement of absolute and relative frequencies in smart sensors. The developed conversion technologies based on the method of the dependent count open the possibilities of the creation of frequency-to-code converters of a new generation.

The accuracy of conversion is one of the most important quality factors for smart sensors that together with reliability characterize their serviceability and high metrological trustiness of obtained results. In traditional methods and frequency-to-code converters it predetermines such metrics of the economic efficiency as dimensions, the weight, the power consumption and the cost price. As a rule, cheaper converters have low accuracy and reliability. The method of the dependent count breaks this link and ensures higher metrics economic efficiency. Among them are:

- The self-adapting mode of measurement for frequency or a period, absolute and relative deviations and ratios. Due to this mode the error of conversion is constant for a wide range of measurand frequencies and the maximal quantization error does not depend on latter.

- Intellectualization of the measurements.

- Programmability of characteristics and functionalities at high metrological trustiness of the obtained results.

- The non-redundant conversion time for all frequencies of the measuring range: from infralow up to high frequencies. Due to this the time of quantization is a variable, set up individually during the quantization for each of frequencies according to the program-specified relative error and thereof is minimal possible.

- The possibility to measure various electrical and non-electrical values transformed beforehand into the frequency with representation of the results into program-specified units.

- The parallelism at synchronous measurements of frequencies practically without the degradation of metrological and dynamic characteristics.

- High efficiency of measurements by super accurate measurement by using an external frequency standard.

- The possibility of creating virtual measuring instruments and systems.

All these are possible due to the simple circuitry and the use of the computational power (microcontroller) as part of the measuring channel.

Due to the method of the dependent count precision and multifunctional smart sensor systems will be more available for customers for the execution of reference measurements at the level of accuracy and stability of the physical constants, as, for example, in the case of the proposed measuring instrument of the magnetic induction based on nuclear magnetic resonance [130].

The method of the dependent count (the method with the non-redundant conversion time) and the method with the non-redundant reference frequency are self-adapting. The first chooses automatically the required conversion time in order to provide the set value of the conversion error. The second method chooses automatically the reference frequency on the set value of the conversion error, thus saving the power consumption in those conditions of the measurement when a precise conversion is not required.

6

SIGNAL PROCESSING IN QUASI-DIGITAL SMART SENSORS

Digital signal-processing techniques are being used in a wide range of industrial and consumer products due to their accuracy and repeatability. According to Texas Instruments digital signal processing is defined as 'The science concerned with representation of signals by sequences of numbers and the subsequent processing of these number sequences' [131]. Processing involves either extracting certain parameters from a signal or transforming it into a form that is more applicable. The digital implementation of signal processing has several distinct advantages:

- It is possible to accomplish many tasks inexpensively that would be either difficult or impossible in the analog domain, for example, Fourier transforms.
- Digital systems are insensitive to environmental changes and component tolerances and ensure predictability and repeatability.
- Reprogrammability features.

Most signal-processing algorithms for quasi-digital smart sensors involve a multiply, divide and an add (subtract) operation which can be written in its general form as Equation (5.26). Before the appearance of embedded microcontrollers and DSP processors, these operations were realized in the quasi-digital domain with frequency–pulse signals. Because a smart sensor uses a microcontroller or a DSP microprocessor in its architecture, it is expedient to perform these operations in the digital domain. However, sometimes in time-critical applications as well as in automatic control, pulse-frequency systems mathematical transformations with pulse-frequency signals are still used. We shall consider some basic transformations peculiar to the frequency-(period)-to-code conversion as well as to signal processing, used for sensor accuracy improvement (quantization error reduction) and to increase sensor noise stability.

6.1 Main Operations in Signal Processing

6.1.1 Adding and subtraction

Often in frequency signal processing it is necessary to subtract one frequency from another or to add two or more frequencies. As the initial information is coming into the adder as continuous sequences of pulses with the frequency $F_i(t)$, proportional to instant values $x_i(t)$, the summation will be reduced to a new sequence of pulses with the frequency $F(t)$ formed according to the following equation:

$$F(t) = k \cdot [F_1(t) + F_2(t)], \tag{6.1}$$

where $F_1(t) = kx_1(t)$, $F_2(t) = kx_2(t)$ is the initial information in a pulse-frequency form; k is the constant coefficient. In its turn, subtraction is reduced to the sequence of pulses with the frequency $F(t)$ formed according to the equation:

$$F(t) = k \cdot [F_1(t) - F_2(t)], \tag{6.2}$$

From the two initial equations for frequency determination according to the method of the dependent count, the following equations of algorithmic transformations and determination of the absolute sum of two frequencies are constructed

$$f_x = f_{x1} + f_{x2} = f_0 \left(\frac{N_{x2}}{N_2} + \frac{N_{x1}}{N_1} \right), \tag{6.3}$$

the absolute sum of two periods

$$T_x = T_{x1} + T_{x2} = T_0 \left(\frac{N_2}{N_{x2}} + \frac{N_1}{N_{x1}} \right). \tag{6.4}$$

Similar equations can be obtained for the determination of an absolute and relative difference of two frequencies (periods), which will differ only in the minus sign in equations (6.3) and (6.4). Besides, knowing the quantization time T_q it is possible to determine the rate of a change in time of absolute and relative sums and differences of periods and frequencies.

Frequency adders and subtraction devices are used as components in complex devices such as frequency multipliers with a positive and negative feedback [132].

6.1.2   Multiplication and division

In frequency measurements, frequency multipliers play the same role as amplifiers of electric signals in amplitude measurements: they increase the sensitivity of measuring devices and extend measuring ranges for smaller values. They are able to improve frequency-to-code converters simultaneously in several directions. First, at a given speed they allow reduction of the quantization error. Second, by a given quantization error it is possible to reduce the time of measurement and, hence, to use the measuring device to control many slowly varied parameters or to reduce dynamic errors by the measurement of one quickly varied parameter. Finally, frequency multipliers can be used for frequency signal unification, allowing the same measuring device to be used with sensors of different output frequencies. Let us consider this in detail.

Despite the fact that frequency multipliers are well known and have been applied in radio engineering and metrology (by phase and time measurements) for a long time, the appearance of frequency-measuring techniques with rather specific requirements to multipliers has resulted not only in the creation of new multipliers, but also in a new understanding of the multiplication itself [4].

The *frequency multiplication* is a conversion of the input electric oscillation with the frequency f_{xi} into the output oscillation with the average frequency:

$$f_{x\text{out}} = k_m \cdot f_{xi},\qquad(6.5)$$

where k_m is the multiplication factor, representing an exact integer number (or in some cases an improper fraction). Thus, the average frequency of a signal is an average crossing by this signal at a certain level (for example, zero) into one side during a time unit. This definition differs from that used in radio engineering and reflects the specificity of frequency-digital devices. Hereby, the multiplication can be reduced to the frequency scaling. This can be realized with the help of widespread frequency dividers and multipliers controlled by the code equivalent of the scale factor k_m. Among a similar class of devices, it is necessary to mark down the so-called binary frequency multipliers as more perspective. A fractional scale multiplier is its main advantage.

The working range of a multiplier, or a band is characterized by the relation of the maximum frequency to the minimum:

$$D = \frac{f_{xi\,\max}}{f_{xi\,\min}},\qquad(6.6)$$

and also its binary logarithm $\log_2 D$ (the range in octaves) or the decimal logarithm $\lg D$ (the range in decades). For input frequencies outside the working range of a multiplier, $\lg D$ does not change the multiplication factor.

A preliminary frequency multiplication is an effective means of increasing the conversion time for low and infralow frequencies. The output frequency of a multiplier exceeds its input frequency k_m times. Then this multiplied frequency will be converted into a code. Hereupon, by a given quantization error the quantization time T_q can be reduced in k_m times, which increases the number of slowly varying parameters that can be measured with the help of one multichannel data acquisition system. By measuring one quickly varying parameter the dynamic error by the given quantization error will be decreased k_m^2 times. Frequency multipliers, using a given quantization time T_q, allow reduction of the quantization error k_m times.

Frequency multipliers unify output signals of frequency sensors and transducers. This is especially important when a variety of sensors are used in a data acquisition system. Thus, it is desirable, that the output frequencies of all, or even parts of, sensors are multiplied by one frequency multiplier. Therefore, the multiplication factor should be able to change over a wide range without the loss of speed. Frequency multipliers can also carry out different functional transformations of the input frequency, thus they may be used for the non-linearity correction of sensor characteristics. At $k_m < 1$ we have a frequency divider.

The main aspects of frequency multipliers are the multiplication factor, the speed and the frequency range. Frequency multipliers used in frequency-to-code converters, should provide first of all a greater multiplication factor, high speed and a wide working frequency range. These requirements are inconsistent. Really, the increase of k_m is accompanied usually by a narrowing of the frequency range, in its turn, the extension of a frequency range controls the speed reduction.

Alongside these three basic requirements for frequency multipliers, there are additional requirements. First of all, functional transformations of the input frequency, etc.

Frequency multipliers with the pulse form of the input and output signal have received the greatest distribution [95]. If the form of the input signal differs, additional forming should be carried out. However, because of noise interferences during pulse forming, the multiplied period is extracted from the input voltage of the sine wave form with some error. The value of this error determines the achievable value of the multiplication factor k_m.

If the sensor signal has a sawtooth form, then counting pulses are formed at some set levels by the input voltage. The constant slope of the sawtooth voltage and the constant actuation level make the multiplier unsuitable for work in a wide frequency range because the pulses' arrangement in time becomes non-uniform by the frequency deviation from the nominal value due to changes of the sawtooth voltage amplitude. Besides, it is possible to lose part of output pulses in a high frequency range.

Using a symmetric triangular form of the input signal, the multiplication is realized by the repeated full-wave rectification. The dc component is subtracted from the input signals, then the full-wave rectification is carried out. After that, we will have the triangular voltage but with double the frequency. The dc component is subtracted again from this signal and full-wave rectification is realized and so on. Such a multiplier does not require any reactive elements and theoretically can have any large multiplication factor in any frequency range by the constant amplitude of the input signal.

If the form of the input signal differs from the symmetric triangular form it is necessary to form preliminary a signal of multiplied frequencies.

For the sine wave form of the input signal, multiplication with spatial coding with the help of the non-linear signal transformation (for example the squaring or cube involution) is most frequently used:

$$\left.\begin{aligned}(V_m \cos \omega t)^2 &= \tfrac{1}{2} V_m^2 \cos 2\omega t + \tfrac{1}{2} V_m^2; \\ (V_m \cos \omega t)^2 &= \tfrac{1}{4} V_m^3 \cos 3\omega t + \tfrac{3}{4} V_m^3 \cos \omega t\end{aligned}\right\} \tag{6.7}$$

The dc component due to the squaring, is eliminated by the appropriate bias. The ac frequency component after the cube involution is eliminated by the subtraction of a certain part of the input signal. Multipliers of this type also suppose the cascade connection of any number of cascades by the not limited frequency range.

Frequency multipliers by the rectangular form of input pulses can be realized most simply. Original structures of such multipliers are described in [133].

6.1.3 Frequency signal unification

Let us consider an example of the multiplier application for frequency signal unification. The large variety of frequency output sensors allows the measurement of various physical and chemical quantities. Depending on the measurand and sensor type the output frequency can vary for a wide range: from fractions of Hz up to several MHz. In multichannel data acquisition systems, with various frequency output sensors, the wide range of input frequencies complicates the use of unified devices which are intended, as a rule, for the input unified frequency signal 4–8 kHz or the standard interface signal 2–22 kHz.

In this case, it is expedient to use devices for preliminary unification of the frequency signal, which play the same role as scaling amplifiers in voltage or current measurements.

The frequency matching device can be constructed on the basis of a digital frequency multiplier, conventional to modern requirements of high accuracy, speed, reliability and stability for a wide range of frequencies. Their multiplication factors can be chosen automatically depending on the ratio between input and unified frequencies.

The circuit of a multirange frequency matching device providing the reduction of a low-frequency range (4–8 kHz) working with a 58-channel multiplexer of frequency signals is shown in Figure 6.1 (Patent No. 798 831, 847 505 (USSR)), [96].

The transformation of the low frequency to this range is carried out by multiplication of a certain factor $k_m = 2^m$, which is set up automatically. For this aim, the period of the input signal is measured in view of the automatic choice of the optimum measuring range, and then the received number is used for forming the output frequency.

The device works in the following way. Each pulse of the input frequency starts the univibrator, developing the short duration pulse. The counter CT1 and the frequency divider are reset by this pulse to the state '0', and the shift register is set '1'. Pulses of the highest clock frequency f_0 come into the counter through the logical circuit AND-OR. After the limiting number N_{max} enters the counter, the output pulse of the multiplexer writes down the number $N_{max}/2$ into the counter. The same pulse enters the clock input of the shift register and shifts the '1' in the register into the single place. The frequency of pulses coming into the counter is decreased two times. The repeated toggle of the logic circuit results in the writing of the number $N_{max}/2$ in the counter and in decreasing clock frequency.

Such a process continues up to the next pulse of the input frequency, which starts the univibrator. The number N_{Tx}, written at this moment into the counter and proportional to the period duration, is rewritten into the third register by the wavefront of the

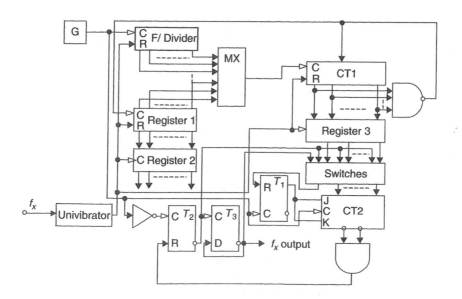

Figure 6.1 Frequency matching device

univibrator's pulse. The number of the chosen range, fixed in the shift register, is rewritten into the second register. Then the counter CT1 and the frequency divider are reset to '0', and the shift register to '1' and the quantization of a new period of the input frequency begins again.

The number N_{Tx} stored in the third register, through switches, is written in the decrement counter and the trigger T1. Pulses of the clock frequency, coming into the decrement counter from the generator G, reduce the number written in the counter. When this number becomes equal to zero, the circuit AND toggles. Its output signal sets up the trigger T2 to '0'. Hereupon, switches are opened and the number N_{Tx} is written in the decrement counter CT2 again. After half a period of the clock frequency the trigger T2 is toggled to '1'.

The subtraction of clock frequency pulses from the number N_{Tx} repeats. The frequency of pulses from the trigger T2 is divided by two with the help of the trigger T3 in order to provide the off-duty factor of multiplied pulses, equal to two.

The decrement counter CT2 has one bit less than the third register. The whole part of the number $N_{Tx}/2$ is written in it through switches. The trigger T1 is intended for registration of the least significant digit of the number N_{Tx}. The switch for its toggle to '0' opens the trigger T3 only once per two switchings of other switches. If the least significant digit of the number N_{Tx} is '1', the trigger T1 is toggled into '0', forming the forbidding signal on (J–K) inputs of the decrement counter. Thus, the first clock frequency pulse after that will not change the condition of the decrement counter but will only toggle the trigger T1 to '1'. After that, the forbidden potential from (J–K) inputs of the counter is removed and the pulse subtraction will continue.

Let us determine the output frequency of the matching device. The number, which is fixed in the third register, equals

$$N_{Tx} = \frac{f_0}{2^m} \cdot T_x, \qquad (6.8)$$

where m is the number of the frequency range.

The duration of two neighbouring cycles between switches toggle is determined in the following way:

$$T_i = T_0 \text{ent} \left\{ \frac{N_{Tx}}{2} \right\}, \qquad (6.9)$$

$$T_{i-1} = T_0 \text{ent} \left\{ \frac{N_{Tx}}{2} \right\} + T_0 \cdot \left(N_{Tx} - 2\text{ent} \left\{ \frac{N_{Tx}}{2} \right\} \right), \qquad (6.10)$$

where T_0 is the period of the clock frequency.

The period of the output frequency T_x equals $T_i + T_{i+1}$, and the output frequency considered with Equation (6.7) is

$$f_{x\_\text{output}} = 2^m \cdot f_x. \qquad (6.11)$$

It is expedient to estimate the error of a frequency-matching device by the instability of the period of the reference frequency, when the maximum value does not exceed one period of the clock frequency. Hence, in order to increase the accuracy, it is necessary to choose a high value of the clock frequency. If the limiting frequency of the electronic components functioning f_{max} is accepted as such a frequency, then at the minimum period $T_{x\,min}$, the maximum value of the relative error will be equal to

$$\delta_{max} = \frac{1}{f_{max} \cdot T_{x\,min}}. \tag{6.12}$$

The counter CT1 and the third register should have the number of bits determined by the following equation:

$$n = \log_2(f_{max} \cdot T_{x\,min}). \tag{6.13}$$

For example, at $f_{max} = 10$ MHz and a unified frequency signal in the 4–8 kHz range, the maximum relative error for a frequency-matching device does not exceed 0.05%. The transition time for the output frequency is one period of the input frequency. The dependence of the multiplication factor on the input frequency is shown in Table 6.1.

The use of the trigger T1 and the decrement counter CT2 allows for the minimum possible error, thus providing the duty-off factor of output pulses equal to two.

The number of the range m of frequencies f_x in the described device is in the second register, and such information is absent from the output frequency signal.

In some cases, the number of the range m can be transferred by the duty-off factor or pulse duration. For this aim, it is necessary to rebuild the unit for forming output pulses. Other parts of a frequency-matching device remain the same.

6.1.4 Derivation and integration

Differentiation and integration are frequently used in an automatic control system. Differentiation is one of the most difficult operations. In general, it can be presented in the following way:

$$F(t) \cong a\frac{dF_1(t)}{dt} \tag{6.14}$$

where $F_1(t) = kx(t)$ is the input pulse frequency signal; $dF_1(t)/dt$ is the first derivation from $F_1(t)$.

Table 6.1 Dependence of multiplication factor on input frequency

Input frequency, Hz	Output frequency, Hz	Frequency multiplication ratio	Code (the number of frequency range m)
<3.9	0	—	—
3.9–7.81	4–8	1024	111
7.81–15.62	4–8	512	110
15.62–31.25	4–8	256	101
31.25–62.5	4–8	128	100
62.5–125	4–8	64	011
125–250	4–8	32	010
250–500	4–8	16	001
500–1000	4–8	8	000
>1000	>8	8	—

In an ideal case, the equation for a derivative of some function of time can be written as the following ratio:

$$F(t)_{\text{ideal}} = \lim_{\Delta t \to 0} \frac{F_1(t + \Delta t) - F(t)}{\Delta t} = \frac{dF_1(t)}{dt}. \qquad (6.15)$$

However, with continuous information, represented in the pulse-frequency form, tending Δt to zero, strictly speaking, loses a physical sense, from the moment when Δt becomes equal to $1/F_1(t)$. Hence, in this case it is possible to speak about tending Δt not to zero, but to some fixed small value $\Delta \tau$, much smaller in comparison with the period T_x. Instead of the infinitely small quantity $dF_1(t)$, it is possible to use the small, but final increment $\Delta F_1(t)$, i.e. in this case, the following ratio will be valid:

$$F(t) = \frac{F_1(t + \Delta t) - F(t)}{\Delta \tau} = \frac{\Delta F_1(t)}{\Delta \tau} \approx k \frac{dF_1(t)}{dt}. \qquad (6.16)$$

Thus, for the pulse frequency signal it is possible to speak only about approximate differentiation of time functions. Thus, the accuracy of differentiation will be higher, if the following inequality is more strictly carried out:

$$\frac{1}{F_1(t)_{\text{min}}} < \Delta \tau \ll T_x. \qquad (6.17)$$

Many measuring tasks, connected with the determination of fluid or gas flows, for example, can be reduced to the integration of some continuous signal. The integration for a pulse frequency signal is reduced to pulse counting with the help of usual or bidirectional counters. The realization of the so-called virtual counters inside a functional-logic architecture of a microcontroller enables such counters to have a large capacity, i.e. the feasible time of integration will be limited in this case only by the internal memory size of a microcontroller. The number of pulses, integrated by the counter, can be expressed by the following dependence:

$$N(\tau) = \int_0^\tau F(t)dt. \qquad (6.18)$$

6.2   Weight Functions, Reducing Quantization Error

Averaging windows allow the error reduction of the average value determination of some signal on its realization, which is limited in time. Averaging windows are weight functions of the finite (duration) impulse response (FIR) low frequency filter with the bandwidth, tending to zero.

The task, which is solved by frequency-to-code converters, can be considered as the dc component determination for some signal containing undesirable pulsations. The counter included in such a converter, counts the number of input pulses during the time T_i (gate time). If the frequency of input pulses is equal to f_x, the number M equal to the product $f_x T_i$, rounded up to the nearest integer value, will be accumulated in the counter. The frequency measurement is shown in Figure 6.2(a). It can be considered as

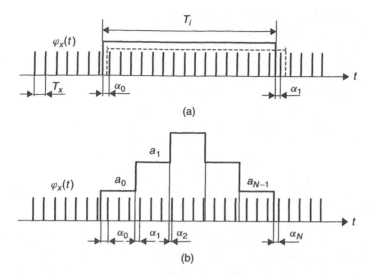

Figure 6.2 Quantization error determination for frequency-to-code converter with Π-shaped
(a) and graded-triangular (b) weight function

the dc component (average value) determination for pulse sequences with the help of
the Π-shaped weight function, or with the help of averaging Dirichlet windows [134].

Let's consider, that pulses, entering the frequency-to-code converter's counter, have
much smaller duration than the period, i.e. can be approximately considered as δ-pulses:

$$\varphi_x(t) = \sum_{n=-\infty}^{\infty} \delta(t - n \cdot T_x),$$ (6.19)

where $T_x = 1/f_x$ is the period of δ-pulses. These pulses enter the counter during a
time limited by the moments t_0 and $t_0 + T_i$. Then the result of the measurement can
be found according to the following formula

$$y = \int_{t_0}^{t_0+T_i} \varphi_x(t)\mathrm{d}t.$$ (6.20)

Let's present a sequence of δ-pulses (Equation 6.19) as the inverse Fourier transform:

$$\varphi_x(t) = f_x \sum_{n=-\infty}^{\infty} e^{j2\pi k f_x t} = f_x + 2 \cdot f_x \sum_{k=1}^{\infty} \cos{(2\pi k f_x t)}.$$ (6.21)

Substituting Equation (6.21) for (6.20), after simple transformation:

$$y = \int_{t_0}^{t_0+T_i} f_x \sum_{k=-\infty}^{\infty} e^{f2\pi k f_x t}\mathrm{d}t$$

$$= f_x T_i + \frac{2}{\pi} \sum_{k=1}^{\infty} \frac{1}{k} \cdot \sin(\pi k f_x T_i) \cos\left[2\pi k f_x \left(t_0 + \frac{T_i}{2}\right)\right].$$ (6.22)

From the received ratio, the measurement result contains the desirable information $f_x T_i$ and the error caused by harmonics with frequencies $k f_x$ in the input signal (Equation 6.21). This error of frequency-to-code converters is known as a quantization error. In reality, during the pulse count the integer number can be received, so the product $f_x T_i$ is rounded to the nearest smaller or greater integer number. In the example in Figure 6.2(a), the product $f_x T_i$ is equal to 16.7. According to this, 16 pulses (the absolute quantization error is -0.7) or 17 pulses (the quantization error is $+0.3$) will enter the counter. Solid and dashed lines showing the Π-shaped weight function in Figure 6.2(a) correspond to these two cases accordingly.

The quantization error is caused by presence of the time interval α_0 and α_1 in Figure 6.2(a) between the wavefront of the Π-shaped weight function and the nearest next pulse of the sequence $\varphi_x(t)$ and between the wavetail of the weight function and the nearest next pulse. We shall consider how the quantization error is connected to the values α_0 and α_1. The frequency-to-code conversion in this case can be presented as the determination of a number of periods T_x, during the time T_i. The number M in the counter is the number of pulses with delimiting periods T_x in the time interval T_i. It is one more than the integer number of periods T_x in the interval T_i. It is obvious, that this time interval equals

$$T_i = T_x \cdot (M - 1) + \alpha_0 + (T_x - \alpha_1). \tag{6.23}$$

As only the value $T_x M$ is taken into account, the absolute quantization error will be equal to

$$\delta = \alpha_0 - \alpha_1. \tag{6.24}$$

Generally, α_0 and α_1 represent the independent random variables distributed according to the uniform distribution law with mean equal to $T_x/2$ and dispersion equal to $T_x^2/12$. The mean of the quantization error $\delta = \alpha_0 - \alpha_1$ is equal to zero, and the dispersion $\delta^2 = \alpha_0^2 + \alpha_1^2$ is equal to $T_x^2/6$. The relative mean-root square quantization error for frequency-to-code converters with the Π-shaped weight function can be determined according to the following equation:

$$\tilde{\gamma}_\Pi = \frac{\tilde{\delta}}{M \cdot T_x} = \frac{1}{\sqrt{6} \cdot f_x T_i}. \tag{6.25}$$

The averaging Dirichlet window (the Π-shaped weight function) is not the best averaging window. In this case, we are dealing with the signal containing a certain set of higher harmonics. Here, the criterion for windows estimation (the greatest amplitude of the amplitude–frequency characteristic lobe) is impossible to use when the window is intended for the reduction of the mean-root square quantization error of frequency-to-code converters.

Weight factors of the optimum graded window reducing the relative mean-root square quantization error, are determined by the ratio:

$$a_n = (n + 1) \cdot (N - n). \tag{6.26}$$

In particular, at $N = 8$ we have $a_0 = a_7 = 8$; $a_1 = a_6 = 14$; $a_2 = a_7 = 18$; $a_3 = a_4 = 20$ (Figure 6.3(a)).

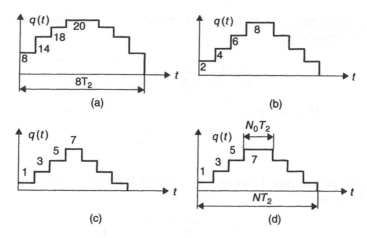

Figure 6.3 Weight functions reducing the quantization error: optimal (a), triangular with even (b) and odd (c) step number, trapezoidal (d)

The absolute quantization error in the case the graded weight function can be calculated according to the equation (Figure 6.2(b)):

$$\delta_q = (\alpha_0 - \alpha_1)a_0 + (\alpha_1 - \alpha_2)a_1 + \cdots + (\alpha_{N-1} - \alpha_N)a_{N-1}$$
$$= \alpha_0 a_0 + \alpha_1(a_1 - a_0) + \alpha_{N-1}(a_{N-1} - a_{N-2}) - \alpha_N a_{N-1}$$
(6.27)

The mean-root square error will be equal to:

$$\tilde{\delta}_q = \frac{T_x}{\sqrt{12}} \sqrt{a_0^2 + \sum_{n=1}^{N-1}(a_n - a_{n-1})^2 + a_{N-1}^2}.$$
(6.28)

The number of pulses, which have been counted by the counter of the converter, can be calculated according to:

$$M = \frac{f_x \cdot T_i}{N} \sum_{n=0}^{N-1} a_n$$
(6.29)

Thus, the relative mean-root square quantization error will be equal to:

$$\tilde{\gamma}_q = \frac{\tilde{\delta}_q}{M \cdot T_x} = \frac{N \cdot \sqrt{a_0^2 + \sum_{n=1}^{N-1}(a_n - a_{n-1})^2 + a_{N-1}^2}}{\sqrt{12} \cdot f_x T_i \sum_{n=0}^{N-1} a_n}$$
(6.30)

The ratio of the error $\tilde{\gamma}_q$ to the error $\tilde{\gamma}_\Pi$, received in the case of the Π-shaped weight function is represented as:

$$V = \frac{\tilde{\gamma}_q}{\tilde{\gamma}_\Pi} = \frac{N \cdot \sqrt{a_0^2 + \sum_{n=1}^{N-1}(a_n - a_{n-1})^2 + a_{N-1}^2}}{\sqrt{2} \cdot \sum_{n=0}^{N-1} a_n} \tag{6.31}$$

For the optimum weight function described by Equation (6.26), it is possible to obtain the following:

$$\tilde{\gamma}_{\text{opt}} = \sqrt{\frac{N}{(N+1) \cdot (N+2)}} \cdot \frac{1}{f_x \cdot T_i} \tag{6.32}$$

$$V_{\text{opt}} = \sqrt{\frac{6 \cdot N}{(N+1) \cdot (N+2)}}. \tag{6.33}$$

As can be seen, the optimum weight function reduces the relative mean-root square quantization error and this reduction increases with the growth of N of the weight function.

The graded-triangular weight function (Figure 6.3(b) and (c)), is close to optimum. For this weight function by the even number N

$$a_n = N + 1 - |N - 1 - 2n|, \tag{6.34}$$

and by the odd N

$$a_n = N - |N - 1 - 2n|. \tag{6.35}$$

By the use of these weight functions in frequency-to-code converters the ratio of the mean-root square quantization error to the mean-root square error of a converter with the Π-shaped weight function will be equal (by the odd and even number of steps N accordingly):

$$V_{\Delta\text{odd}} = \sqrt{\frac{8 \cdot N^2}{(N+1)^3}}; \tag{6.36}$$

$$V_{\Delta\text{even}} = \frac{2\sqrt{2} \cdot N - 3}{N} \tag{6.37}$$

The trapezoidal weight function provides slightly better results than the triangular weight function. It represents the triangular weight function calculated according to Equations (6.34) or (6.35), but with a cut-off point (Figure 6.3(d)). The analysis shows that the best result is reached when the ratio of the base of trapezoid to the top is approximately equal to three ($N/N_0 \approx 3$).

According to Equation (6.33) at $N \gg 1$ and by transition from the Π-shaped to the optimum weight function the relative mean-root square quantization error is reduced V_{opt} times:

$$V_{opt} \approx \sqrt{\frac{6}{N}} \tag{6.38}$$

The same ratio

$$V = \frac{\tilde{\gamma}_q}{\tilde{\gamma}_\Pi} \tag{6.39}$$

for triangular weight functions can be calculated on the basis of Equations (6.36) and (6.37):

$$V_\Delta \approx \sqrt{\frac{8}{N}} \tag{6.40}$$

For the trapezoidal weight function at $N/N_0 \approx 3$ and $N \gg 1$ it is possible to calculate the ratio:

$$V_{trapezium} \approx \sqrt{\frac{6.75}{N}}. \tag{6.41}$$

Thus, in all considered cases, the more steps N that the weight function contains the less the quantization error will be. Thus, triangular and trapezoidal weight functions give the quantization error, only in $\sqrt{8/6} \approx 1.15$ and in $\sqrt{6.75/6} \approx 1.06$ times more, than the optimal weight function. Taking into account that triangular and trapezoidal weight functions can be realized (by hardware or software) a slightly easier than the optimum weight function, we can arrive at the conclusion that these weight functions are expedient for quantization error reduction in frequency-to-code converters.

In the classical direct counting method for frequency-to-code conversion, the conversion time T_i is casually located in relation to pulses, with frequency f_x measured (Figure 6.2(a)). However, sometimes the wavefront of the time interval T_i is synchronized with one of the specified pulses. In this case, there is no need to reproduce the weight function as a whole. It is enough to reproduce only the second half of the triangular, trapezoidal or optimum weight function.

In this section, we have considered weight functions which reduce the quantization error in frequency-to-code converters. The weight method of averaging frequency (period) conversion results is also useful, when the transformation is carried out in the low and infralow frequency range and under the influence of industrial noises.

The quantization error reduction can be reached also by using weight functions in period, phase shift converters as well as in integrating analog-to-digital converters with a pulse-frequency or time-pulse intermediate conversion. Therefore, using the weight function in the phase-shift-to-code converter, the limiting absolute error for Dirichlet window averaging is 3.6°, and for the triangular weight function $-0.1°$, for a frequency of 50.5 Hz and conversion time 1 s. Modern DSP microprocessors allow the use of more complex weight functions, for example, synthesized with trigonometrical components.

Summary

Signal processing from the quasi-digital smart sensor point of view can be performed in a frequency-time domain using pulse-frequency signals as well as in a digital domain after frequency (period)-to-code conversion.

The use of optimum weight functions (for example, triangular) as advanced signal processing increases the accuracy of smart sensors (by reducing the quantization error) and increases the noise stability.

7

DIGITAL OUTPUT SMART SENSORS WITH SOFTWARE-CONTROLLED PERFORMANCES AND FUNCTIONAL CAPABILITIES

Digital systems are being ever-increasingly used for measurement and control applications. However, all the variables in the 'real world' which sensors are used to measure (such as temperature, pressure, flow or light intensity) are analog in their physical nature: an element is therefore always needed to link the analog environment to the digital system. This usually also means that signals from sensors must be appropriately modified, so as to be made suitable for conversion into a digital data format.

The interface from the analog domain to the digital domain can be a mystifying design problem. The hardware design and the software must operate together to produce a complete, usable design. It is especially true for the smart sensor design. Here the hardware and the software are needed to implement the bridge between the system analog signals and the digital signals.

The use of frequency-to-code converters based on the so-called *program-oriented conversion methods (PCM)* in combination with frequency output sensors and transducers of electric and non-electric quantities by the creation of smart sensors with embedded microcontrollers was considered earlier (Figures 1.2, 3.7), and seems to promise a lot. In this case, the conversion error is determined by the sensor's accuracy. The hardware or the chip area can be reduced to the minimum possible (the microcontroller core and peripherals).

Not all conventional conversion algorithms are suitable for use in program-oriented conversion methods. Sometimes it is necessary to modify the algorithm or to create in essence a new one taking into account the corresponding microcontroller's capabilities. The basic feature of program-oriented conversion methods is that the programmable computing power is directly included into a converter as part of a measuring circuit and takes part in conversion, i.e. some transformations in measuring procedures executed in a digital form are included. The transformation of the frequency-to-code converter's structure from the hardware realization to the software essentially changes conversion methods and functional possibilities of such devices. Such program-oriented conversion methods for frequency-time domain parameters were proposed in 1988 for the first time and published in 1989–95 with the aim to minimize the hardware (the chip area) and create small-sized, highly reliable smart sensors and sensor microsystems which have minimum power consumption, high metrological performances and self-adapting capabilities.

A definition of the program-oriented conversion method was given in [135]: A PCM is the processor algorithm of measurement, incarnated in the functional-logic structure of a computer or a microcontroller through the software.

For program realization of conversion methods all elementary measuring procedures of the processor algorithm are carried out by the program and the measuring circuit is realized at a virtual level inside the functional-logic architecture of the reprogrammable computing capacity. Due to the dualism of realization of elementary operations of the measuring procedure inherent in the architecture of modern microcontrollers, the realization of a whole class of PCMs, which are based only on one conversion method, for example, the ratiometric counting method, is possible. Thus, such a method of measurement represents the basis for all possible PCMs. In fact, the PCM is a program model of hardware realization of a measuring circuit and can be unequivocally set up by the concrete basis and the processor algorithm of its realization.

Depending on the program realization there are two types of program model construction: compilating and interpreting. PCMs are procedures with complex functional-logic structures containing critical time-dependent pieces of programs so that their algorithms, according to the specified classification are constructed according to the compilating way of the program realization by which for every PCM the specific program is realized.

The variety of PCMs demands further development of their classification, taking into account detailed attributes for the ordering of existing and the development of new PCMs, for example, as in the work [99] concerning methods of measurement used in frequency-to-code converters based on a hardwired logic.

It is possible to characterize a PCM by certain generalizing attributes, which are peculiar to each PCM based on any basis and method of measurement. According to the program realization of frequency and time references, PCMs can be subdivided into methods:

- with program delays formed
- with time intervals formed by means of built-in timer/counters (T/C).

Depending on the realization of the events counting algorithm, PCMs can be subdivided into methods:

- with counting on polling (synchronous or asynchronous)
- with counting on interruptions (single-level or multilevel)
- with counting with the help of the timer/counters.

7.1 Program-Oriented Conversion Methods Based on Ratiometric Counting Technique

Let's consider the realization of the program-oriented method for the frequency-to-code conversion based on the ratiometric counting technique. Algorithms of measurements, based on this method, demand a high multisequencing degree of elementary measuring procedures of processor algorithms. For a long time this was a constraint on such PCMs on the basis of traditional microprocessor systems. However, modern microcontrollers have well-developed functional-logic architectures and instruction sets focused on an effective realization of input/output procedures, providing, thus, a multiway PCM realization based on the ratiometric counting method. High efficiency of PCMs by design should be provided by careful analysis of processor algorithms' structures, the use of resources of microcontrollers' architectures and correct determination of the software part, which requires the optimization.

In the design, the PCM can be represented as a set of software components of a high-level hierarchy. The latter, in its turn, consists of elementary low-level software components. It is an elementary procedure realized at the virtual level. So, for PCM realization based on the ratiometric counting method a concurrent execution of four elementary procedures is necessary: the first gate time T_{01} forming; the second gate time T_{02} forming, the wavefront and the wavetail of which are strictly synchronized with pulses of the input frequency f_x; the counting of reference f_0 and measurand f_x frequencies (Figure 5.2).

The decomposition of the PCM into elementary measuring procedures and variants of its possible software realizations are shown in Table 7.1.

In order to generate the set of possible PCM realizations, it is expedient to analyse all possible realizations of these operations, given by the functional-logic opportunities of microcontrollers.

The first gate time T_{01} can be formed by a built-in timer/counter or by a software time delay constructed with the help of nested iterations and the combination of precisely trimmed delays of smaller duration.

Ranges of possible delays by the timer/counter for microcontrollers of MCS (Intel) and MSP430 (Texas Instruments) families are adduced in Table 7.2. Here f_{osc} is the frequency of the crystal oscillator; n is the number of bits of the timer/counter. Delays of greater duration can be realized by construction of cycles with timer/counter reloads.

Table 7.1 Variants of possible software realizations for measuring procedures

Elementary procedures		Variants of realization			
		By delay	By T/C	By interrupt	By pooling
T_{01}	f_x	•	•	—	—
N_1		—	•	•	•
T_{02}		—	—	•	•
N_2 / f_0		—	•	•	•

Table 7.2 Ranges for possible delays for timers/counters

Type of μK	Minimum time delay	Maximum time delay
MCS-51	$12/f_{osc}$	$12/f_{osc} \times 2^n$
MCS-251	$4/f_{osc}$	$4/f_{osc} \times 2^n$
MCS–96/196	$4/f_{osc}$	$4/f_{osc} \times 2^n$
MSP430CXX (at MCLK $= 1.048$ MHz)	1.5×10^{-5} s	2 s

By using a timer/counter for the forming of time intervals, there is an error Δt_{int} caused by the time of instruction execution necessary for the control transfer to the interrupt vector of the internal interruption and the 'stop' instruction for a timer/counter. It is a systematic error and can be corrected.

From time delays forming, the minimum possible duration is determined by the minimum microcontroller's instruction execution time. As a rule, it is the NOP (no operation) instruction. Its execution time is equal as a rule to one machine cycle. The inefficient use of microcontroller's resources is the main disadvantage of programming time delay forming. The forming of second gate time T_{02} is often carried out by means of the logic analysis of events. The counting of pulses f_x and f_0 can be realized by one of three possible ways: by software asynchronous polling; by interruptions and with the help of T/C. Each of these ways brings the restrictions connected with additional errors of measurement, which must be taken into account by the metrological analysis of the developed PCM, and, whenever it is possible, must be corrected.

The realization of a PCM with counting of events with the help of program asynchronous polling becomes simpler due to the microcontroller's conditional jump instruction according to the presence of a high logic level '1' in the microcontroller's inputs. However, in order to prevent pulse losses of the input frequency f_x as well as a false reading, the maximum converted frequency and the pulse τ_x duration should be chosen from the following system of inequalities:

$$\left. \begin{array}{c} T_{x\,max} \geq n \cdot \tau_{cycle} \\ \tau_{Jump} \leq \tau_x < n \cdot \tau_{cycle} \end{array} \right\}, \tag{7.1}$$

where $T_{x\,max}$ is the period of the converted frequency; τ_{cycle} is the duration of the machine cycle; n is the number of machine cycles necessary for realization of the polling program; τ_{Jump} is the execution time for the conditional jump instruction according to the presence of a high logic level '1' in the microcontroller's inputs.

The maximum converted frequency for a PCM with counting of events by interruptions is limited by the ratio

$$T_{x\,max} \geq T_{INT}, \tag{7.2}$$

where T_{INT} is the execution time of the interrupt subroutine. Thus, it is necessary to provide the presence of an active level of a signal of the external interruption during the time:

$$T_{INT} \geq t_{clc} \cdot n, \tag{7.3}$$

where n is the number determined for each type of microcontroller.

The maximum converted frequency for a PCM with the counting of events by means of built-in timer/counters is limited only by electronic components, i.e. by the maximum frequency, which can be counted by the timers/counters. The minimum duration of a signal on the timer/counter inputs is regulated by specifications on the microcontroller. Such a PCM allows frequency-to-code converters with the greatest possible converted frequency f_x and the minimum quantization error. Maximum possible frequencies on timer/counter's inputs for various microcontroller families are adduced in Table 7.3.

The minimum converted frequency $f_{x\,min}$ is practically unlimited and determined by the maximum virtual counter capacity, which is determined in its turn, by the capacity of all free general-purpose registers and all accessible sizes of the internal microcontroller's RAM:

$$f_{x\,min} = \frac{1}{N_{2\,max} \cdot T_{clc}}, \tag{7.4}$$

where $N_{2\,max} = 2^k = const$; k is the capacity of the virtual counter, realized in general-purpose registers and/or in the RAM for the reference frequency pulse f_0 calculation. The graph of the function $f_{x\,min}(f_0, N_2)$ is shown in Figure 7.1.

The use of microcontrollers with an architecture containing two or more built-in timer/counters, for example, microcontrollers 8X52, 8XC5X, 8XL5X, 8XC51FX, 8XL51FX and 8XC51GB (Intel) expands the high range of converted frequencies for all specified advantages of the PCM.

If a third timer/counter is used for forming the reference time interval T_{01}, converter speed will be increased. Thus, quasi-pipeline data processing can be realized. One more advantage of the microcontroller architecture with three timer/counters is an

Table 7.3 Maximum possible frequency on timer/counters inputs for various microcontroller families

Type of μK	Maximum frequency on T/C input
MCS-51	$f_{osc}/24$
MCS-251	$f_{osc}/8$
MCS–96/196	$f_{osc}/4$
MSP430C33x (for MCLK)	3.8 MHz

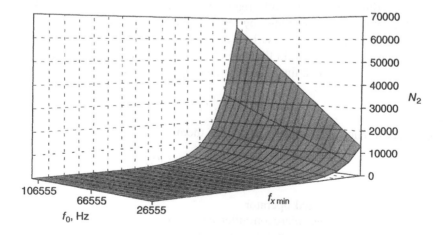

Figure 7.1 Graph of function $f_{x\,min}(f_0, N_2)$

opportunity to use the third timer/counter in the software-controlled frequency output Clock-Out mode. In this mode, the meander with the programmed frequency of pulses is generated on the T2 (P1.0) output. This mode is useful for forming the reference frequency f_0 pulses. In this case, it is not necessary to use the preliminary frequency divider in the converter. The output frequency on the T2 depends on the frequency of the crystal oscillator and the number reloaded into the timer from registers RCAP2H, RCAP2L and can be calculated according to the formula:

$$F_{\text{clock-out}} = \frac{F_{\text{osc}}}{4 \cdot (65536 - \text{RCAP2H,RCAP2L})} \tag{7.5}$$

The increase of the reference frequency f_0 up to $F_{\text{clc}}/4$ is possible due to the PCM realization based on microcontrollers containing a programmable counter array (PCA). It is a good internal device for measurement of the period, the pulse duration, phases differences, etc., in five channels at the same time. Thus, there is a possibility to create converters with increased accuracy due to quantization error reduction.

It is necessary to take into account that not only methodical errors caused by processing algorithm properties, but also the computing capacity built-in to the measuring circuit as well as programming style influence the conversion result. Hence, in order to reduce systematic and random errors of the measurement, it is expedient to build the PCM algorithm in a rational way from the metrological efficiency point of view, i.e. software optimization with the help of the morphological analysis according to the chosen criteria—the minimum of these components.

According to the metrological analysis, the conversion error includes components caused by software properties that have been written in a low-level computer language (assembler). This error can be minimized by using an optimal PCM.

The optimal operator is determined by the equation

$$A^*_{\text{opt}} = \underset{A \in \{A\}_{\text{accept}}}{\operatorname{argmin}} \; r_2(A_k x_k, A x_k) \tag{7.6}$$

where $r_2(A_k x_k, A x_k)$ is the non-identity error, caused by the difference between the measuring algorithm and the actually realized algorithm by the software; $\{A\}_{\text{accept}}$ is the allowable set of operators limited by requirements, from which the A^*_{opt} is chosen. One of the *sine qua non* to the class $\{A\}_{\text{opt}}$ of allowable operators is that the number of possible realizations is finite. In the case of the discrete optimal choice of the PCM, the set $\{A\}_{\text{accept}}$ is set up as a final set of known operators with known parameters:

$$\{A\}_{\text{accept}} = \{A_i(c_i^*)\}, \; i = \overline{1, k}, \tag{7.7}$$

It is necessary to choose the best variant from them

$$i^*_{\text{opt}} = \underset{i \in \{\overline{1,k}\}}{\operatorname{argmin}} \; r(A_k x_k, A_i(c_i^*)) \tag{7.8}$$

where $A^*_{i\,\text{opt}}(c_i^*)$ is the required operator.

Such a task arises by the microcontroller choice as well as at the synthesis of the optimum PCM from the finite number of possible elementary measuring procedures of the processor algorithm.

At the final design stage the value of the algorithmic error with its maximum permissible value Δ_2 is compared according to the following rule

$$\left.\begin{array}{c} r_2 < \Delta_2 \Rightarrow 1 \\ r_2 \geq \Delta_2 \Rightarrow 0 \end{array}\right\} \tag{7.9}$$

The inequality $r_2 \geq \Delta_2$ indicates:

1. For Equation (7.6)—within the frame of given restrictions the requirement $r_2 < \Delta_2$ is not valid, inasmuch as the value r_2 is characterized by the limited (potential) quality of the PCM. Assuming that

$$\lim_{\nabla\Gamma \to \infty} r^{(2)} = r_1, \tag{7.10}$$

   where r_1 is the theoretical error, we conclude that it is necessary to increase the value Γ, i.e. the reconsideration of given data.

2. For Equation case (7.8)—it is necessary to chose another microcontroller or to create the PCM based on another basis (Equation 7.6), for example, based on the method of the dependent count.

Recommendations on the basis of the morphological analysis for designing the minimum PCM hardware for frequency-to-code converters on the basis of a microcontroller, are based on the main metrological features such as:

- The open character of frequency-to-code converter functionalities.

- An opportunity to exchange accuracy to speed and conversely (self-adapting capabilities) or extension of a measuring range up to high frequencies.

- Practically unlimited range of low and infralow converted frequencies.

- A high degree of multisequencing operations of the conversion algorithm.

- The essential complication of smart processor algorithms (adaptation, correction, calibration, self-diagnostics; statistical processing etc.).

- The growth of the relative density and complication of the structure of the methodical and, first of all, algorithmic component of the error of measurements.

Due to the functional-logical capabilities of modern microcontrollers and dualism of realizations for the main operations of the conversion algorithm, the 10–18 realizations of software components at a high level are possible on average. From set theory, the program-oriented method of conversion can be represented as the union of p disjoint sets N of software realizations of elementary operations of the i-th conversion algorithm:

$$S^* = S_1 \cup S_2 \cup S_3 \ldots \cup S_p (S_1 \cap S_2 \cap S_3 \ldots S_p \neq 0), \tag{7.11}$$

where $S_1, S_2, S_3, \ldots, Sp$ includes $n_1, n_2, n_3, \ldots, n_p$ elements accordingly. The determination of a general number of possible conversion methods (the first design stage) for the particular basis and a microcontroller is the combinatory task. For a generation of PCM variants with the help of a choice of necessary combinations from sets $S_1, S_2, S_3, \ldots, Sp$, it is necessary to combine each choice of elements from the set S_1 with each choice of elements from $S_2, S_3, \ldots, S_p$. Similarly, each choice of elements from $S_2, S_3, \ldots, S_p$ should be combined with a choice of elements from $S_1, S_3, \ldots, S_p$, from $S_1, S_2, \ldots, S_p$ and $S_1, S_2, S_3, \ldots, S_{p-1}$ accordingly. Therefore the number of elements of the set R^* (i.e., the tuple from k elements of the set S^*), from which $k_1 \in S_1, k_2 \in S_2, k_3 \in S_3, \ldots, k_p \in S_p$ is equal to:

$$V^n = \{V_i^n\} = \prod_{i=1}^{p} C_{N1}^{ki} - q = C_{N1}^{k1} \cdot C_{N2}^{k2} \cdot C_{N3}^{k3} \cdot \ldots \cdot C_{Np-1}^{kp-1} \cdot C_{Np}^{kp} - q, \qquad (7.12)$$

where q is the number of incompatible realizations of the PCM, i.e. operations that cannot be executed within the frame of a certain microcontroller's architecture simultaneously. So for example, the pulse counting of f_0 and f_x can be realized with the help of the timer/counters, however, in the frame of a one-timer/counter microcontroller's architecture it is impossible to carry out these two operations simultaneously.

For PCM realization on the basis of the ratiometric counting technique it is necessary, as described earlier, to realize three elementary procedures: to form the first gate time T_{01} and count pulses of the unknown f_x and reference f_0 frequencies. Then, the general number of alternative, possible in principle, variants of the PCM in view of incompatible operations for the one-timer microcontroller's architecture (one timer/counter and one input of external interruptions), is determined as:

$$V^n = \{V_3^n\} = C_3^1 \cdot C_3^1 \cdot C_2^1 - 8 = 10 \qquad (7.13)$$

For microcontrollers of the base configuration with two timer/counters:

$$V^n = \{V_3^n\} = C_3^1 \cdot C_3^1 \cdot C_2^1 - 1 = 17, \qquad (7.14)$$

and for microcontrollers of the extended configuration with three timer/counters:

$$V^n = \{V_3^n\} = C_3^1 \cdot C_3^1 \cdot C_2^1 = 18 \qquad (7.15)$$

The algorithmic structure of the PCM in many respects depends on the used basis (conversion method) and the microcontroller type. The choice of the optimum PCM, from the metrological criterion of the efficiency point of view, among all varieties of allowable realizations, is not a trivial task. The successful choice of the required PCM in many respects is determined by designer experience. The choice of the optimum discrete PCM is carried out with the help of the design methodology for reusable software components of smart sensors [136], which will be described in the following section.

7.2 Design Methodology for Program-Oriented Conversion Methods

The discrete choice of optimum alternative variants of the PCM in the finite set of allowable realizations represents a characteristic task of vector synthesis (optimization).

By a discrete choice, the set of strongly admissible systems forms the given discrete finite set M_{sa} of points in the n-dimensional Euclidean space R^m of the characteristic. The task of synthesis consists in the choice of such a point $\overline{a}^* \in M_{sa}$ from this set, which has the best value of the vector of characteristic

$$K = <k_1, \ldots, k_m> \longrightarrow K_{\min(\max)}, \tag{7.16}$$

that is, the best in the sense of the chosen criterion of preference, and satisfying all sets $D = \{C, O_s, QV, O_k\}$ of initial data, where $C = \{C_1, \ldots, C_p\}$ is the set of conditions; $O_s = \{O_{s1}, \ldots, O_{sq}\}$ is the set of restrictions on the structure and parameters of the designed frequency-to-code converter (the hard constraint on the used microcontroller type and the conversion method); $QV = <k_1, \ldots, k_m>$ is the structure of the vector of characteristic of the PCM; $O_k = O_{k1}, \ldots, O_{kr}$ is the set of restrictions, imposed on parameters of a quality [137].

From all strictly allowable PCMs, we shall choose the optimum PCM that has the best value of the vector of characteristic K. Hence, one of the major procedures of vector synthesis is the choice of the optimization criterion K.

For further description of the design methodology of optimum PCMs, it is necessary to take advantage of following definitions [137]:

Definition 1. *The system S satisfying the set $\{C, O_s\}$ of the initial data is referred to as allowable.*

Definition 2. *The allowable system satisfying the set of restrictions O_k is referred to as strictly allowable, i.e. satisfying the whole set $D = \{C, O_s, VQ, O_k\}$.*

In general, the processor algorithm of the PCM is estimated by some set of parameters of efficiency

$$K = \{k_i\}, i = \overline{1, m}, \tag{7.17}$$

and it is usual that $|K| = m > 1$.

For successful synthesis of the optimum PCM, it is necessary to choose such a subset $\tilde{K} \in K$, which most fully characterizes various properties of the frequency-to-code converter in view of its application (smart sensor features) and requirements. On the other hand, for simplification of the synthesis of the frequency-to-code converter based on the PCM, it is desirable to limit the subset $\tilde{K}$ whenever possible by parameters that estimate in the greatest degree the ability of the device to carry out its functions from the application point of view.

As the structure of the frequency-to-code converter based on a certain microcontroller core is fixed (rigid restrictions on the structure), it is not necessary to compare various PCMs according to the complexity S. On the other hand, on the basis of the microcontroller type and conversion method V_n PCM variants can be realized according to Equation (7.12); important specific and essential characteristics by comparison of alternative PCM variants are the quantization error $\delta(\%)$ and the maximum possible converted frequency $f_{x\,\max}$ (Hz).

As the PCM supposes various algorithms of realization $\{A_{\text{accept}}\}$ from the set of allowable variants of the realization M_a, another essential characteristic by PCM comparison can be the memory sizes V_{ROM} (byte) and the power consumption P. These last two parameters are rather essential for one-chip smart sensor realization.

With provision for the negative ingredient for parameters of quality, the optimization criterion can be written, for example, as follows:

$$\nabla_i K_i = \min_{\{B_{ij}\}} F_i(\delta_{ij}, T_{xij}, V_{ROMij}, P_{ij}) \qquad (7.18)$$

at $T_{ij} \le T_{ij\,accept}$, where $T_{ij\,accept}$ is the allowable conversion time; $\{B_{ij}\}$ is the set of allowable variants for the ith PCM, and $\delta_{ij}, T_{xij}, V_{ROMij}, P_{ij}$ are factors of the jth allowable variant of the ith PCM; $T_{xij} = 1/f_{xij}$.

For the forming of the integrated parameter of the quality K_i, it is expedient to take advantage of the weight-average geometrical complex parameter, which is written as follows

$$\prod_{i=1}^{m} \left(\frac{k_i^{min}}{k_i}\right)^{v_i}, \qquad (7.19)$$

where k_i^{min} is the minimum value of appropriate parameters of quality on all allowable variants of the PCM; v_i is the normalized weight factor of proportionality for appropriate parameters, and

$$\nabla_i v_i \ge 0 \text{ and } \sum_{i=1}^{m} v_i = 1 (i = \overline{1, m}). \qquad (7.20)$$

Then the integrated criterion of the PCM efficiency can be submitted as a multiplicate loss function:

$$K_{ij} = \left(\frac{\delta_{i\,min}}{\delta_{ij}}\right)^{v_i\delta} \cdot \left(\frac{T_{xi\,min}}{T_{ij}}\right)^{v_{iT}} \cdot \left(\frac{V_{ROMi\,min}}{V_{ROMij}}\right)^{v_{iROM}} \cdot \left(\frac{P_{i\,min}}{P_{ij}}\right)^{Pi}, \qquad (7.21)$$

where $\delta_{i\,min}, T_{xi\,min}, V_{ROMi\,min}, P_{i\,min}$ are the minimum values of appropriate parameters of quality on all allowable variants of the ith PCM; $v_i\delta, v_{iT}, v_{iROM}, v_{iP}$ are the weight factors of priority for appropriate parameters of quality; T_{xi} is the period of the converted frequency. Thus

$$\nabla_i 0 \le \{v_i\delta, v_{iT}, v_{iROM}, v_{iP}\} \qquad (7.22)$$

and

$$(v_{i\delta} + v_{iT} + v_{iROM} + v_{iP}) = 1 \qquad (7.23)$$

In Equation (7.21), instead of $f_{xi\,max}$ the value $T_{xi\,min}$ is used to reduce this parameter of quality to the standard kind.

The criterion function according to Equation (7.21) characterizes the set of effective solutions well enough. In other words, the maximization of this function gives the effective solution for the initial multicriterial task. In the contrast to other methods using weight factors, these factors are well interpreted in the given scalarization. They depend on the desirable value of criterion: if the desirable level is closer to the optimum value, the criterion is more important and the weight factor has the greater value. In its turn, the allowable variant of the PCM realization is that for which $T_{ij} \le T_{ij\,accept}, i = \overline{1, m}$ is true. The optimal variant from all those allowable will be the variant that has the maximal value K_{ij}, and $K_{ij\,max} = 1$. Thus, K_{ij} is the non-linear function of normalized dimensionless parameters $\{k_i\}$. Depending on the weight factors of parameters of quality, which are included in the integrated criterion K_{ij}, it is possible to choose a PCM, optimizing it according to one of the following parameters: the minimal period of converted frequencies, the quantization error, the ROM memory size, the power consumption, etc.

Equation (7.18), in view of the criterion resulting function (Equation 7.21), represents a non-linear discrete task of optimum designing.

As follows from different references devoted to methods of vector optimization, a reliable solution of the synthesis task can be obtained only as a result of the combination of various optimality criteria, in particular, an unconditional and conditional criterion of preference.

In view of the above, it is possible to allocate the following stages of the discrete choice of the optimal PCM for frequency–time parameters:

1. The analysis of the PCM's algorithm and restrictions on its realization, the forming and ordering of the set M_{sa} of strictly allowable PCMs (catalogue compiling of alternative, possible in principle, variants of software realizations).

2. The set partition M_{sa} into the set of worse M_w and non-worse M_{nw} systems with the help of the unconditional criterion of preference (finding of the left lower boundary).

3. The use of the conditional criterion of preference for searching of the optimum PCM in the set M_{nw} when the set of non-worse systems is nondegenerated.

4. The analysis of synthesis results on stability and definiteness.

Further, it is expedient to provide remarks and practical recommendations for the execution of these design stages.

The preparative stage.   Before this synthesis stage, it is necessary to choose the PCM basis (as considered above, it is expedient to choose any of the advanced methods for the frequency-to-code conversion, for example, the ratiometric counting method) and the type of the embedded microcontroller or the microcontroller core. In general, the task of microcontroller choice for the frequency-to-code converter and optimization estimations of functional-logic features of the microcontroller architecture, are tasks of the structural synthesis.

The set of given microcontrollers m is the open set, which is extended by new microcontrollers or standard libraries of cells in CAD tools. Therefore, the optimal microcontroller for the use in frequency-to-code converters is the microcontroller (or the microcontroller core) that provides PCM realization within the frame of given restrictions at minimum expense:

$$S_0 = \min S_i; \quad i = \overline{1, \varepsilon};$$

$$T_{ij} \leq T_{ij\,accept} \tag{7.24}$$

where ε is the maximum number of different microcontrollers.

Stage 1.   In order to obtain the solution closest to the optimal, the PCM must be decomposed into a number of elementary procedures by the method of decomposition: forming the gate time T_{01}, the pulse counting of f_x, f_0 etc., and then the synthesis method through the analysis of allowable variants $\{v_i^n\}$ is used.

The enumeration of possible alternative variants of the PCM in the finite set M_{sa} depends on the discrete choice task. Because the set M_{sa} of the PCM is foreseeable and finite ($N_A = 10$–18) the algorithm of full enumeration and the exclusion of unpromising variants based on the method of the consecutive analysis of variants is used in order to find the optimum PCM from a small, known number of variants. In order to use this method it is necessary to compose a catalogue of all PCM variants. The common number of alternative variants $v_i^n \in M_{sa}$ is determined according to Equations (7.12–7.15), and their generation is carried out according to Equation (7.11). In order not to miss alternatives, the set M_{sa} should be preliminary ordered.

The morphological matrix is frequently used for the construction of a catalogue of possible program-oriented conversion methods from the high-level software components. However, this matrix does not reflect the specificity of decomposition of program-oriented methods. For this aim, it is more convenient to take advantage of the morphological tensor $\omega_A(a, b, c)$, which is created based on the functional-logical analysis of developed program-oriented methods. Methods of the software realization of the three basic elementary procedures of the algorithm (Table 7.1) are selected as the most important criteria. In this tensor, the maximum value of the index a is determined by the number of possible realizations of the pulse count f_x, b by the number of realizations of the pulse count f_0, and c by the number of possible variants of forming the gate time T_{01}. The numerical values of indexes and realizations of elementary operations, appropriate to these values, are shown in Table 7.4.

The tensor's element X_{abc} is a Boolean variable. Its value is equal to '1' by the choice of the program-oriented conversion method, otherwise it is equal to '0'. The condition of alternatives can be expressed as follows:

$$\sum_{i=1}^{n} X_{abc} = 1, \; i = \overline{1, n} \qquad (7.25)$$

Further description of the technique for a discrete choice of optimum program-oriented conversion methods is provided by an example of the sensor microsystems synthesis based on widespread MCS-51 (Intel) microcontroller families. Let the microcontroller contain only one timer/counter and be able to service one interrupt from one external source.

Table 7.4   Numerical values of indexes and elementary operations realizations

Index (a, b, c)	Operation	Description
a_i, b_j	1	Count of f_x or/and f_0 by polling
	2	Count of f_x or/and f_0 by timer/counter
	3	Count of f_x or/and f_0 by interrupt
c_k	1	Software delay
	2	Delay with the help of timer/counter

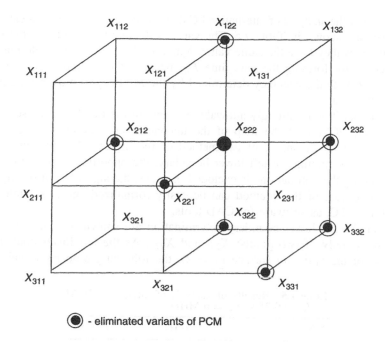

Figure 7.2 Morphological block of possible alternative variants of a PCM

The morphological block of possible alternative variants of program-oriented conversion methods, from which incompatible q variants for the given microcontroller operations are eliminated, is shown in Figure 7.2.

The operation can be presented by the indexes ratio of tensor's elements:

$$\begin{cases} a = b = c \vee a = b \vee b = c \vee a = c \\ a, b, c \geq 2 \end{cases} \tag{7.26}$$

Further dimension reduction of morphological space is carried out by the argumentation of the integrated criterion of efficiency.

Stage 2. As the set $\{v_i^n\}$ of strictly allowable PCMs, generated according to Equations (7.12–7.15) is nondegenerated and finite $(1 < I < \infty)$, the following theorem [137] is valid:

Theorem. *If M_{sa} is a finite set, the set of non-worse systems M_{nw} is the nonempty set.*

In order to simplify the M_{nw} determination, it is expedient to exclude minor parameters of quality from the integrated criterion of efficiency at the initial stage. First of all, these are V_{ROMij} and P_{ij} (Equation (7.18)) as they do not influence the PCM metrological efficiency. Their maximum values are determined at the preparative stage of the synthesis and limited by functional-logic features of the chosen microcontroller. Thus, the synthesis task of the optimum PCM has been reduced to the choice of the optimum PCM according to the maximum of a metrological criterion of efficiency, determined only by two parameters of quality: δ_{ij} and T_{xij}.

The results of the analysis of alternative PCMs according to the metrological criterion of efficiency are shown in Table 7.5. The variants distinguished only by the index c, can be joined, as they have the same values δ and $f_x(T_x)$. These values do not depend on the method for forming the gate time T_{01}, because the reference time interval is always formed with the accuracy determined by the accuracy of a quartz generator of the microcontroller.

For the set partition of strongly allowable variants of the PCM into the set of worse and non-worse PCM with the help of the unconditional criterion of preference, it is expedient to take advantage of the method of rectangles [137]. In comparison to other methods for finding the left lower boundary, the opportunity to lose non-worse PCMs, including the optimal one is excluded. Besides, the algorithm of the equivalent analytical procedure of this method can be easily formalized and is suitable for the realization on a PC, as software for CAD tools.

Stages for the left lower boundary determination are shown in Figure 7.3. In this case it consists only of two points: X_{231} and X_{321}. As the left lower boundary is a special case of the optimum surface $(m = 2)$, the following property is valid for it:

Table 7.5   Results of analysis of alternative PCM (at $T_{01} = 0.25$ s, $f_{clc} = 6$ MHz)

PCM	δ_x (%)	f_x (kHz)	T_x (s)
X_{112}	0.014	22.2	4.5×10^{-5}
X_{211}	0.014	133.3	0.75×10^{-5}
X_{312}	0.014	33.3	3.0×10^{-5}
X_{121}	0.003	22.2	4.5×10^{-5}
X_{321}	0.003	33.3	3.0×10^{-5}
X_{132}	0.012	22.2	4.5×10^{-5}
X_{231}	0.012	133.3	0.75×10^{-5}

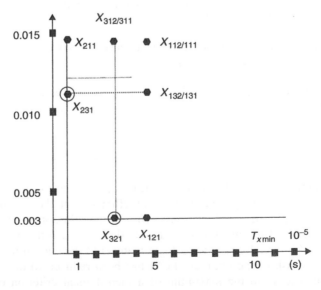

Figure 7.3   Stages of left lower boundary determination

on the whole expansion of the left lower boundary the dependence $\delta = f(T_{x\,min})$ and consequently, the $T_{x\,min} = \phi(\delta)$, has the monotonely decreasing character.

It is now necessary to take advantage of the conditional criterion of preference in order to finish the synthesis procedure.

Stage 3.   Taking into account excluded minor parameters of quality the resulting criterion function will be:

$$K_{ij} = \left(\frac{\delta_{i\,min}}{\delta_{ij}}\right)^{v_i\delta} \cdot \left(\frac{T_{xi\,min}}{T_{ij}}\right)^{v_{iT}},\qquad(7.27)$$

and the formulation of the synthesis task can be written in the following way

$$\left.\begin{array}{l} K_p = f_p(K_1, \ldots, K_i, \ldots, K_m = \max\limits_{S\in M_a} \\[2mm] K_i = K_i(S), i = \overline{1, m} \\[2mm] K_i \le K_{im}, i = \overline{1, m} \end{array}\right\}\qquad(7.28)$$

where K_{im} is the value of the factor K_i that is the maximum possible. Because the nondegenerated set M_{nw} contains only two PCMs, one of which has the least value of the quantization error $\delta_{i\,min}$, and the other the minimum value of the converted period $T_{xi\,min}$, the integrated quality factor for the variant X_{231} is reduced into the formula

$$K_{ij} = 0.25^{v_i\delta},\qquad(7.29)$$

and for the variant X_{321} into the formula

$$K_{ij} = 0.167^{v_{iT}}\qquad(7.30)$$

A more optimum variant is X_{321} with the values $K_{ij} = 0.5$ against 0.41 accordingly by equivalent requirements for the PCM accuracy and speed (weight factors of priority for quality indexes are equal, i.e. $v_{i\delta} = v_{iT} = 0.5$). Plots of functions (7.29) and (7.30) with allowance for relations $v_{i\delta} + v_{iT} = 1$ are shown in Figure 7.4. These dependencies

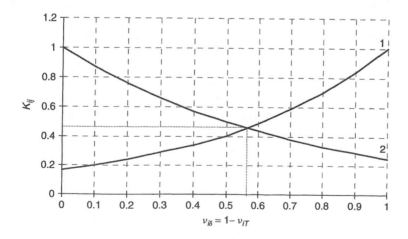

Figure 7.4   Graph of function $K_{ij} = f(v_{i\delta})$ (1) and $K_{ij} = f(v_{iT})$ (2)

can be used during the synthesis of the PCM, facilitating a choice of the PCM according to requirements of speed or accuracy.

Stage 4.   It is obvious that the PCM synthesis is completed if the following two conditions are valid:

1. The solution obtained as a result of the synthesis, is determined enough and precise, i.e. the PCM class, satisfying all formulated initial data, including the optimal criterion, is not too wide.

2. The received solution (class PCM) is steady enough by the variation in reasonable limits of initial data, including preference relations.

Really, the nondegenerated set M_{nw} contains two PCMs and the result of the synthesis will be similar using the criterion of efficiency (Equation (7.21)) instead of (Equation (7.27)) as well as by taking into consideration variants distinguished by gate time forming the T_{01}.

The given technique of the discrete PCM choice is applicable also to microcontroller architectures containing two or more built-in timer/counters. Thus, the generation of possible alternative variants of the PCM is made according to Equations (7.14), (7.15) accordingly. The morphological block of possible PCM variants will be similar to that shown in Figure 7.2. It is necessary to eliminate only one variant, satisfying the following condition

$$a = b = c, \quad a, b, c \geq 2 \tag{7.31}$$

i.e. the variant X_{222} in case of the microcontroller architecture with two built-in timer/counters. In the case of improved architectures with three or more timer/counters, there are no variants of the PCM with incompatible operations. In the first case, the use of the unconditional criterion of preference at the second synthesis stage results into the unique solution X_{221} (the set M_{nw} is degenerated), in the second case, into the solution $X_{221/222}$. Thus, based on the conditional criterion of preference according to the resulting criterion function (Equation (7.21)), preference should be given to the variant X_{222}, as this requires less program memory size for its realization.

Thus, the task of the choice of the optimal PCM can be reduced in general to the non-linear discrete task of optimal designing. This technique covers almost all functional-logic architectures of modern microcontrollers, suitable for PCM realization based on classical and advanced methods of the frequency (period)-to-code conversion for its use in different smart sensors. The technique is well formalized and allows algorithms of all stages of the morphological synthesis, thus providing an opportunity for automatization of procedures of the discrete choice of PCM and the creation on this basis of appropriate software for modern CAD tools [138].

7.2.1   Example

The control flow chart of an optimum PCM, with the microcontroller family MCS-51 (Intel), is shown in Figure 7.5 and its software in Figure 7.6 respectively. As microcontrollers of this family contain two–three 16-bit timer/counters, it is expedient to realize

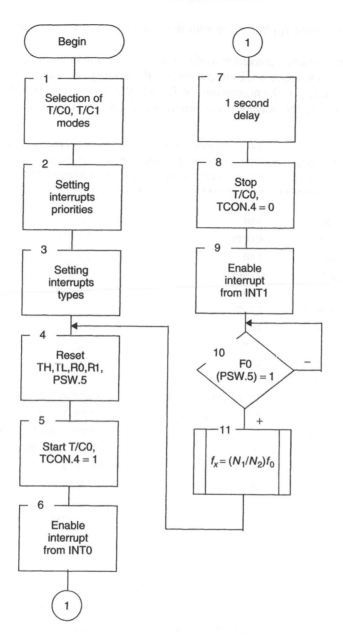

Figure 7.5 Control flow chart of optimal PCM based on MCS-51 microcontrollers

the PCM in the following way. Pulses of frequencies f_x and f_0 are counted simultaneously by two timer/counters, and the T_{01} form with the help of the software delay or by the third timer/counter (if it is in the microcontroller). Timer/counters overflows cause internal interruptions, which are counted by virtual counters based on general-purpose registers R0 and R. Thus, the greatest possible converted frequency and the minimum quantization error are reached. The reference frequency f_0 is formed by the frequency divider. The gate time T_{02} is formed with the help of two level interruptions on inputs INT0, INT1.

The frequency-to-code converter works as follows. At the beginning of the process, operation modes for T/C0, T/C1 are chosen and the priority of external interruptions is established: for INT1/0–the maximum; for T/C0/1–the lowest (blocks 1,2). Then types of interruptions (on the wavetail) are programmed (block 3), all registers of virtual

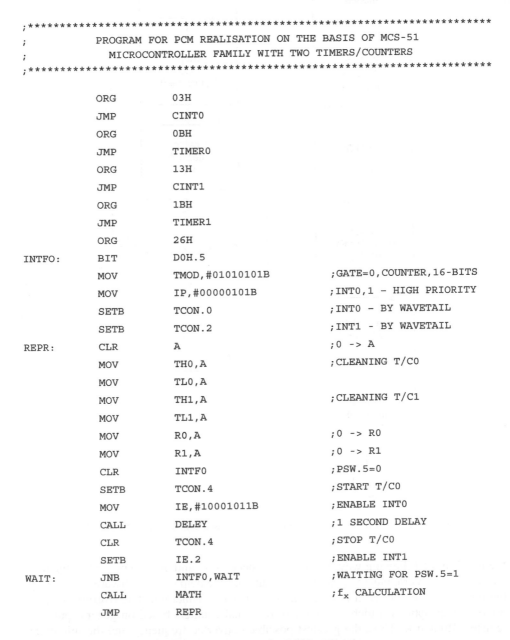

```
;***********************************************************************
;           PROGRAM FOR PCM REALISATION ON THE BASIS OF MCS-51
;              MICROCONTROLLER FAMILY WITH TWO TIMERS/COUNTERS
;***********************************************************************

            ORG       03H
            JMP       CINT0
            ORG       0BH
            JMP       TIMER0
            ORG       13H
            JMP       CINT1
            ORG       1BH
            JMP       TIMER1
            ORG       26H
INTFO:      BIT       D0H.5
            MOV       TMOD,#01010101B       ;GATE=0,COUNTER,16-BITS
            MOV       IP,#00000101B         ;INT0,1 - HIGH PRIORITY
            SETB      TCON.0                ;INT0 - BY WAVETAIL
            SETB      TCON.2                ;INT1 - BY WAVETAIL
REPR:       CLR       A                     ;0 -> A
            MOV       TH0,A                 ;CLEANING T/C0
            MOV       TL0,A
            MOV       TH1,A                 ;CLEANING T/C1
            MOV       TL1,A
            MOV       R0,A                  ;0 -> R0
            MOV       R1,A                  ;0 -> R1
            CLR       INTF0                 ;PSW.5=0
            SETB      TCON.4                ;START T/C0
            MOV       IE,#10001011B         ;ENABLE INT0
            CALL      DELEY                 ;1 SECOND DELAY
            CLR       TCON.4                ;STOP T/C0
            SETB      IE.2                  ;ENABLE INT1
WAIT:       JNB       INTF0,WAIT            ;WAITING FOR PSW.5=1
            CALL      MATH                  ;fx CALCULATION
            JMP       REPR
```

Figure 7.6 Program for PCM realization

```
CINT0:      SETB        TCON.6                      ;START T/C1
            CLR         IE.0                        ;DISABLE INT0
            RETI


CINT1:      CLR         TCON.6                      ;STOP T/C1
            CLR         IE.2                        ;DISABLE INT1
            SETB        INTF0                       ;SETTING PSW.5=1
            RETI


TIMER0:     INC         R0
            RETI


TIMER1:     INC         R1
            RETI
```

Figure 7.6   (*continued*)

counters of N_1, N_2 and the user's bit F0 in the fifth bit PSW.5 are reset (block 4). Then the timer/counter T/C0 for pulse f_x count is started and the interruption from INT0 is enabled (blocks 5,6) After that, the time delay $T_{01} = 1$ s (block 7) is formed using software. When this time finishes, the T/C0 is stopped (block 8), the interruption from INT1 is enabled (the block 9) and the program waits for the last pulse f_x, to finish a measuring cycle (block 10). Then the subroutine for the frequency f_x calculation (block 11) is executed and the measuring cycle is repeated again.

The interruption subroutine from INT0 starts T/C1 for counting pulses of the reference frequency f_0 and forbids the interruption on the input INT0. In its turn, the interruption subroutine from INT1 stops T/C1, forbids the interruption INT1 and sets up the user bit in '1'. Interruption subroutines from the timer/counters overflow increment virtual counters.

The PCM used in such a converter has a high degree of the elementary operations concurrency of the measuring algorithm. The high conversion frequency range is limited only by the greatest possible value of frequency, which the timer/counters can count.

7.3   Adaptive PCM with Increased Speed

For PCM realization of frequency–time parameters of signals with the pulse counting of frequencies f_x by interruptions and forming the gate time T_{01} by the method of the nested iteration and a combination of precisely trimmed delays of the smaller duration, there is a problem connected with T_{01} increasing because of the summation of times of external τ_{INT} and internal τ_{TC} interruption subroutine executions with the gate time

T_{01}. In general, the total conversion time without taking into account mathematical processing, is determined according to the following formula

$$T_{\text{conversion}} = \tau_{\text{INT}} \cdot N_1 + \tau_{\text{TC}} \cdot N_m + T_{01} \qquad (7.32)$$

Taking into account the equation for N_1, after transformations we have:

$$T_{\text{conversion}} = \text{ent}\,\{T_{01} \cdot f_x\} \cdot \tau_{\text{INT}} + \tau_{\text{TC}} \cdot N_m + T_{01} = \tau_{\text{delay}} + T_{01} \qquad (7.33)$$

where N_1 is the number of pulses f_x, arriving at the interruption input during the time T_{01}; N_m is the number of internal interruptions on timer/counter overflows. For the high frequency range, the value τ_{delay} becomes more or equal to T_{01}, thus increasing the conversion time and the dynamic error.

The use of adaptive PCMs avoids the loss of speed. By the algorithmic adaptation, the structure of the PCM's processor algorithm is purposefully varied during the frequency-to-code conversion. In general, the controlling factor can be presented as

$$U = \langle A \rangle, \qquad (7.34)$$

where A are the factors, which help to change the algorithmic structure.

As the algorithmic structure of the PCM is characterized by the stable conjunction of elementary procedures, the use of such an adaptation optimizes the structure of the processor algorithm. The task of adaptation can be formalized by the following way:

$$Q < A > \longrightarrow \min; \quad \min \Rightarrow A^*; A \in E_A, \qquad (7.35)$$

where Q is the minimum criterion (in this case, the minimum $T_{\text{conversion}}$); E_A is the set of allowable algorithmic structures A; A^* is the optimal algorithm.

With the aim of time τ_{INT} compensation, which can be calculated according to the formula

$$\tau_{\text{INT}} = \sum_{i=1}^{n_k} p_i \tau_i, \qquad (7.36)$$

where p_i is the instruction with the execution time τ_i; n_k is the number of instructions of the interruption subroutine, the time delay program forming the gate time T_{01} is modified during the frequency-to-code conversion. Thus, the interruption subroutine for external interruptions from pulses f_x should contain at least one decrement instruction for the microcontroller's register DEC <reg>. The number written in this general-purpose register is the controlled loop variable, realizing the time delay equal to the time of the subroutine execution. In its turn, this cycle is included in the program body forming the gate time T_{01}. With such a structure of the PCM's algorithm, each subroutine execution automatically reduces the time T_{01} for the τ_{INT} value and the gate time T_{01} remains constant in all ranges of converted frequencies. The PCM adaptation is realized in parallel to the frequency-to-code conversion on the basis of the current information about the frequency f_x, without the use of the aprioristic information about the measurand. The graph of the function $T_{\text{conversion}} = \phi(f_x)$ for the conventional and adapting PCM is shown in Figure 7.7. The piece of the program forming the gate time

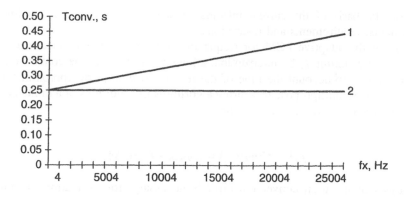

Figure 7.7   Graph of function $T_{\text{conversion}} = \phi(f_x)$ for conventional (1) and adapting (2) PCMs

Figure 7.8   Piece of program forming gate time T_{01}

T_{01}, which meets the requirements of adaptation is shown in Figure 7.8. Thus, the looping time is $\tau_{\text{cycle}} = \tau_{\text{INT}}$.

Apart from the compensation of the subroutine execution times of the interruptions processing, the PCM algorithmic adaptation modifies interrupt vectors during the frequency-to-code conversion. This property of the adaptive PCM allows the realization of multifunctional converters, functioning according to complex measuring processor algorithms in conditions of a limited number of interrupt vectors.

In this case, instead of the unconditional jump instruction to the address of the interruptions subroutine, required by the protocol of the interruptions processing, the indirect jump instruction to the address of the virtual interrupt vector is situated. According to the required algorithm, the interruptions subroutine modifies the virtual interrupt vector at the end of the interrupt processing. Due to this, the following interrupt initializes the execution of another subroutine, which in its turn, can modify the virtual interrupt vector again, initializing the previous or a new interruptions subroutine and so on.

The method of the PCM adaptation with the use of virtual interrupt vectors is effective for both external (hardware) and internal (software) interrupts. Such a protocol of the interruptions service is expedient to use, for example, for PCMs of phase shifts-to-code converters.

In multifunctional converters based on adaptive PCMs, the established set of measuring functions $\{FM_i\}$ is executed with the help of $\{P_{ni}\}$ interruptions subroutines due to which, the set of algorithms $\{AM_i\}$ is realized. The concrete algorithm is chosen on the basis of the current information about the measurand, conditions of measurements, requirements and restrictions.

The use of the adaptive PCM for frequency–time parameters of electrical signals with the pulse counting f_x by interruption, allows an increase in the conversion speed (without taking into account the time of the result processing) by approximately two times and realizes multipurpose measuring algorithms in conditions of a limited number of interrupt vectors in the microcontroller.

7.4   Error Analysis of PCM

The analysis of the total conversion error is necessary, for estimation of dominant components bringing the greatest contribution to the resulting error, for revealing specific errors inherent only to program-oriented conversion methods for the frequency-to-code conversion and also for the local optimization task definition, the solution of which reduces these dominant components.

The general structure of the conversion error for the frequency-to-code converter based on the ratiometric counting method is shown in Figure 7.9. For analytical equations describing errors of synthesized optimum PCMs, it is necessary to investigate separate components of this error. In view of the features of the algorithm execution, it is expedient to choose a structure of the total error, which permits its factor-to-factor research. For this aim, having calculated the total differential of Equation (5.4) and having finite increments and relative units, we obtain the relative conversion error, which can be submitted as:

$$\delta_q = \frac{\Delta f_x}{f_x} = \frac{\Delta N_1}{f_x \cdot T_{01}} + \frac{\Delta f_0}{f_0} - \frac{\Delta N_2}{f_0 \cdot T_{01}}, \tag{7.37}$$

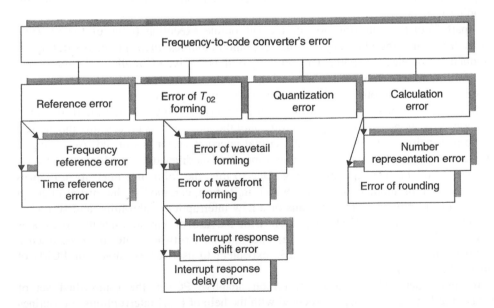

Figure 7.9   General structure of conversion error

where $\Delta f_0/f_0$ is the relative error of reference; $\Delta N_1/f_x \cdot T_{01}$ and $\Delta N_2/f_0 \cdot T_{01}$ are the relative quantization errors. The first component of the quantization error is minimized by the choice of conversion time, multiple to the period of converted frequencies and the use of $f_0 = f_{max}$ for the given microcontroller. The conversion error will thus be equal to

$$\delta_q = \frac{\Delta f_{x\,max}}{f_{max}} \pm \frac{1}{f_{max}} \cdot T_{01}, \qquad (7.38)$$

Frequently, the second component quantization error is determined according to the formula for engineering calculations (Equation (5.7)). A more exact mathematical model (Equation (5.10)) of the given component has been described earlier in Chapter 5.

7.4.1 Reference error

The reference error is represented by two components: the time reference error by the forming of the gate time T_{01} with the help of the software delay and the frequency reference error f_0, by the frequency division or generation with the help of the timer/counter, working in the mode of the programmable frequency output 'Clock-Out'. The reference error is practically identified with the error caused by instability of a reference-frequency source (a quartz generator), inherent to any microcontroller. Change of quartz generator parameters is determined by the influence of a large number of random factors and consequently is a random variable distributed according to the central limiting theorem under the normal distribution law.

The maximum accuracy of frequency-to-code conversion is determined by the frequency stability of the built-in quartz-crystal generator. The system clock generator is a 'measurement standard' in these converters.

The temperature instability of the quartz-crystal generator is one of the main components of the parametric instability, which has the greatest density in comparison with other components. As known, the frequency deviation of the non-temperature-compensated crystal oscillator from the nominal due to the temperature change is $(1–50) \times 10^{-6}$ in the temperature interval $(-55 - +125\,°C)$ and in the frequency range 5–50 MHz. In its turn, modern program-oriented methods of frequency-to-code conversion are able to obtain the quantization error, commensurable with the temperature instability of the references. Therefore, knowledge about the more exact value of the measurement standard uncertainty is a necessary condition for designing accurate frequency-to-code converters for smart frequency–time-domain sensors.

To obtain the exact value of the reference accuracy is possible only by experiment. Experimental researches of quartz-crystal generators have a large complexity because of considerable amount of time consumed on the temperature characterization. Due to a large scatter, appropriate to the real-temperature characteristics of quartz-crystal generators, the experiment assumes a very important place in the design and manufacture of accurate smart sensors. The aim of this section is to acquaint the reader with the technique of experimental researches of the reference error in order to be able to repeat such experiments for any type of microcontrollers.

The research of quartz-crystal generators in the working temperature interval with the aim to determine the temperature instability of frequency is common for all varieties of quartz generators. The temperature instability of frequency in the whole working temperature interval from t_{min} up to t_{max} is determined as [139]:

$$\frac{\Delta f}{f} = \frac{2 \cdot (f_{max} - f_{min})}{f_{max} + f_{min}} \tag{7.39}$$

where f_{max} and f_{min} are maximum and minimum values of the generator's frequencies in the working temperature range.

The objectives of the experiment were [140]: (i) to determine the temperature instability of frequency and time references of the embedded microcontroller; (ii) to determine the correlation factor of their errors; (iii) to determine distribution laws for the temperature uncertainty. As an experimental sample the one-chip microcontroller from the MCS-51 (Intel) microcontroller family was selected, because of its very wide distribution due to the usability of its functional-logical architecture.

According to the algorithm of the method used for the frequency-to-code conversion, frequency and time references were realized based on the built-in quartz-crystal generator. The frequency reference was obtained in the following way. The synchronization signal of the internal system clock generator of the microcontroller through the output ALE was applied to the input of the external frequency divider by three. Then the signal $f_{ALE/3}$ was applied to the input of the built-in microcontroller's timers/counters. The 6 MHz resonator RG-05 was used as a quartz-crystal oscillator. The time reference was obtained in the following way. The required time interval was formed with the help of the delay subroutine. This subroutine was realized by the method of nested program cycles and combinations of precisely set up delays with smaller duration. The schematic diagram of the experimental set up is shown in Figure 7.10. For the

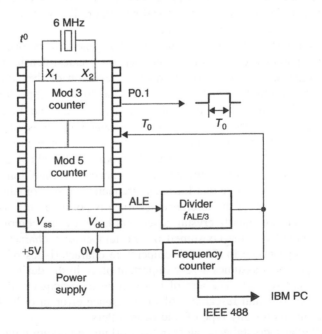

Figure 7.10 Schematic diagram of experimental set up

determination of the temperature instability of frequency the quartz-crystal oscillator was cooled up to a temperature equal to the minimum temperature from the working temperature interval and was then soaked in this point for up to 25–30 minutes. After that, values of frequency on the output ALE and the duration of the time interval on the output P0.1 were measured. The frequency measurements were repeated for remaining points of the working temperature interval through 5–10 °C in all working ranges from t_{min} up to t_{max}. The time of a normal temperature cycle with the temperature overfall of 50 °C should not be less than 15 minutes.

The researched errors are random variables, therefore statistical processing of 60 values was carried out to obtain authentic eventual results. The abnormality of values in the sample was checked with the help of the Shovene criterion. The processing of experimental data was carried out on an IBM PC-compatible computer. The accuracy of measurements is characterized by the following dimensions: the root-mean-square error of measurements σ, the probability error of measurements υ and the average error of measurements η. In this case

$$\sigma = \pm \sqrt{\frac{\sum \varepsilon^2}{n \cdot (n-1)}}, \tag{7.40}$$

where n is the number of measurements; ε is the deviation of separate measurements from their arithmetical average;

$$\upsilon = \pm \frac{2}{3} \cdot \sigma \tag{7.41}$$

$$\eta = \pm 0.8 \cdot \sigma \tag{7.42}$$

Finally, the results of measurements can be written as

$$f_x = \frac{\sum f_i}{n} \pm \sqrt{\frac{\sum \varepsilon^2}{n \cdot (n-1)}} \tag{7.43}$$

The experimental research of the correlation dependence for the temperature uncertainty of measures was carried out by the simultaneous measurement of frequency on the output $f_{ALE/3}$ of the external divider and the time interval T_0 on the output P0.1. The temperature of the quartz-crystal oscillator varied from $+18$ °C up to $+100$ °C with consequent cooling. The research of the correlation dependence was carried out by the change of the supply voltage V_{dd} in the range from $+4.7$ up to $+5.25$ V.

Experimental results for errors of frequency and time references are shown in Figure 7.11 and 7.12 respectively. Frequency changes of an output signal of the quartz oscillator during the time interval from 1 ms up to 10 s are stipulated by the thermal noise of a quartz oscillator and elements of the circuit as well as by fluctuations of other parameters. For a short average time, the main destabilizing factor is the thermal noise of the quartz oscillator. Histograms of their distributions are shown in Figures 7.13 and 7.14. The check of distribution laws according to the χ^2 criterion has shown that distribution laws of errors of measures are close to the Gaussian distribution law.

Although the frequency and time references are realized with the help of the same internal system clock generator of the microcontroller, the temperature uncertainty of the frequency reference is higher. It is stipulated that the required reference frequency

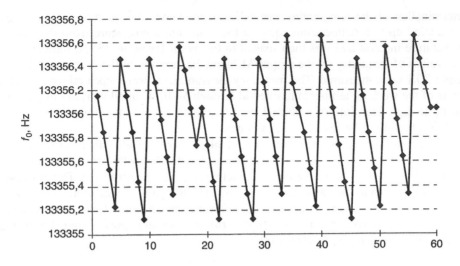

Figure 7.11   Frequency reference error at $t = 20\,^{\circ}\text{C}$

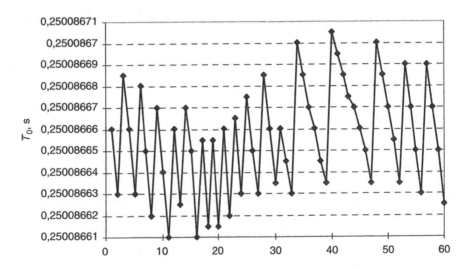

Figure 7.12   Time reference error at $t = 20\,^{\circ}\text{C}$

f_0 is formed by two internal and one external frequency dividers, introducing contributions to the resulting error of the frequency measure.

The thermal instability of the frequency for the quartz-crystal generator is $\Delta f/f = 11.5 \times 10^{-6}$. Dimensions of the measurement accuracy are shown in Table 7.6.

The experimental researches of the temperature uncertainty for frequency and time references realized on the basis of the built-in microcontroller's quartz-crystal generator determined the extreme accuracy of the 'measurement standard', which cannot be exceeded for the given technical implementation independently on the method of measurement (conversion) used. The maximum temperature uncertainty of the

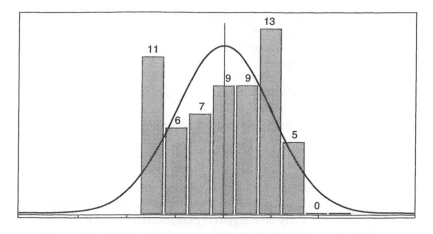

Figure 7.13 Histograms of distributions for frequency reference error

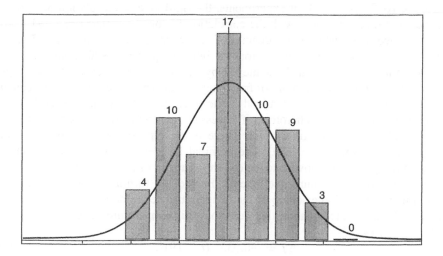

Figure 7.14 Histograms of distributions for time reference error

Table 7.6 Dimensions of measurement accuracy at $P = 97\%$

	Frequency reference	Time reference
Result of measurement	$133\,355.892 \pm 0.468$ Hz	$(250\,086.655 \pm 0.0246) \times 10^{-6}$ s
υ	± 0.312	$\pm 0.0164 \times 10^{-6}$
η	± 0.374	$\pm 0.0197 \times 10^{-6}$
Dispersion	0.219	0.0006

frequency reference is $\pm 11.5 \times 10^{-6}$, and the maximum temperature uncertainty of the time reference is $\pm 0.38 \times 10^{-6}$.

Two components of the reference error considered above are caused by the same reason (frequency instability of the built-in quartz generator) and, hence, are strongly correlated. By the summation of these components, it is necessary to take into account the correlation factor ρ, reflecting that errors of time and frequency references are rigidly and negatively correlated. Generally, it is expedient to present all components by mean-square deviations in order to eliminate the influence of distribution laws from deformation at the error summation:

$$\sigma_{To} = \frac{\delta_{To\,max}}{3} \quad \text{and} \quad \sigma_{fo} = \frac{\delta_{fo\,max}}{3}, \tag{7.44}$$

then the sum of correlated errors is determined according to the formula

$$\sigma_{\Sigma_m} = \sqrt{\sigma_{To}^2 - 2\sigma_{To}\sigma_{fo}\rho + \sigma_{fo}^2} \tag{7.45}$$

If both summable error components are distributed according to the uniform distribution law, the resulting distribution will be trapezoidal. As these errors in these experimental investigations are rigidly and negatively correlated, to find the total error of the measure they should be added algebraically with allowance for the correlation factor ρ. Then the total error of the 'measurement standard' in the microcontroller is equal to $\pm 10.9 \times 10^{-6}$. As a result of experiments, the rigid inverse correlation of errors for frequency and time references with the correlation factor $\rho = -0.981 \pm 1.291 \times 10^{-6}$ was determined. Correlation dependencies of reference are shown in Figure 7.15.

Further improvement of stability for the built-in quartz-crystal generator of the embedded microcontroller can be achieved by the use of the oven-controlled crystal oscillator. The temperature stability of frequency can also be increased by the frequency control of quartz generators, as realized in the microcontroller of the MSP430 (Texas Instruments) family [129]. Therefore, if necessary, the temperature behaviour of the crystal can also be taken into account. There is the typical dependence of a crystal in relation to its temperature [129]. The nominal frequency is preset at one temperature t_0 (turning point); above and below this temperature the frequency is always lower (a negative temperature coefficient). Beside the turning point, the frequency deviation increases with the square of the temperature deviation (-0.035 ppm/$^\circ C^2$ for example). The quadratic equation that describes this temperature behaviour is approximately ($t_0 = +19\,^\circ C$):

$$\Delta f = -0.035 \cdot (t - 19)^2, \tag{7.46}$$

where Δf is the frequency deviation in ppm; t is the crystal temperature in $^\circ C$.

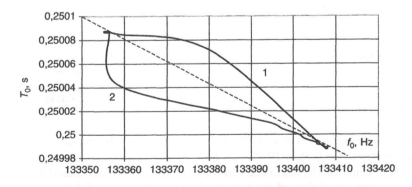

Figure 7.15 Correlation dependencies of reference at heating (2) and cooling (1)

Using the above equation the crystal temperature is measured and the frequency deviation computed every hour. These deviations are added until an accumulated deviation of one second is reached: the counter for seconds is then incremented by one and one second is subtracted from the accumulated deviation, leaving the remainder in the accumulation register.

7.4.2   Calculation error

Inclusion of the processor into a converter measuring circuit and numerical measuring transformations into a measuring procedure changes the structure of the total error. Components caused by numerical measuring transformations appear and the necessity of metrological analysis in connection with conversion algorithm complication arises. It is particularly valid for smart sensors, in which short number bit (8-, 16-bit) microcontrollers are used. On the other hand, there is no necessity essentially to overestimate the requirement for the calculation accuracy in comparison with those errors, which are introduced by other components. It is necessary to assume that the calculation errors could be neglected.

The class of typical computing procedures includes the multiplication and division of multibyte numbers (8-, 16-, 32-, 64-bits) as well as the binary-to-binary-coded decimal code conversion and conversely. Generally, by the PCM realization based on the ratiometric counting method, the calculation of converted frequencies or periods are made according to Equations (5.4) and (5.11) respectively. In general, the limited word capacity of the microcontroller results in two kinds of errors: the numeration error and rounding error. If for the numeration Xn_0 bits are used, the numeration error of X is equal to half of the weight of the lowest-order digit and can be determined as

$$\sigma_{\text{Number\_Representation\_Error}} = \tfrac{1}{2} k^{-n_o}, \tag{7.47}$$

where k is the base of the system of numeration.

The truncation error arises as a result of the word-length reduction after multiplication, division as well as the algebraic summation in view of scaling variables. The same error exists by the operand right shifts.

It is necessary to differentiate between the simple truncation (rejection) and the rounding truncation. The accuracy of the first operation is much lower because the mean is not equal to zero, however its realization is much easier. These components of the error calculation, its characteristics and calculation technique have been described in many references. However, the general theory needs an additional specification with reference to PCM realizations.

In order to increase the computing speed and reduce dynamic errors, the operands realizing dependences (Equations (5.4) and (5.11)) should be used in a fixed point form. Numeration ranges for N_1, N_2 are determined by the range of converted frequencies $[f_{x\,\text{min}}, \ldots f_{x\,\text{max}}]$:

$$N_{1\,\text{max}} = \text{ent}\{T_{01} \cdot f_{x\,\text{max}}\} \tag{7.48}$$

$$N_{2\,\text{max}} = \text{ent}\left\{ \frac{N_1 \cdot f_o}{f_{x\,\text{min}}} \right\} \tag{7.49}$$

Thus, the non-redundant bit number n_0, used for the presentation of a number, is achieved. Multibyte operands are stored in packed format in the microcontroller's memory.

Arithmetic subroutines operate with binary, integer, signless operands. The word length of operands should result from the minimum condition of the error arising owing to the word-length limitation. So, for multiplication, it is expedient to use the multiplication method by the low-order digit ahead with the shift to the right of the partial product. This method works according to the exact multiplication scheme. For multibyte division, it is expedient to use the division method with the shift to the left and the residue restoration. Moreover, by the execution of the division subroutine, it is necessary to check if the denominator equals zero and to process the 'divide by zero' error. The secondary effect of these actions is an opportunity to determine the failure in the reference frequency channel. So, the equality $N_2 = 0$ testifies about the problem with the pulse propagation in the frequency divider $\rightarrow$ timer/counter's input path.

Because the microcontroller's data memory, whose size determines the possible length of operands and results, is used for operands and results storage the accuracy of arithmetic operations is not limited by the word length of the microcontroller, and $\delta_{calc} \ll \delta_q$.

Generally, in Equation (5.4), the reference frequency f_0 is constant and does not belong to the set of integers Z. Then, with the aim of using the integer arithmetic to increase the computing operation accuracy, it is expedient to use scaling by the multiplication of the numerator and the denominator to scale the factor k_m. The use of the multiplication according to the exact scheme with the required word length by which the multiplication result of three n_x-, n_y- and n_z- bit numbers has $n_x + n_y + n_z$ bits, and the certain execution order for mathematical operations with the same priority (three multiplications, then the division), minimizes the calculation error. Thus, the truncation error appears, only at the last stage of calculations which is determined by the number of bits necessary for the representation of the remainder of division.

In view of scaling, the final calculations of frequency should be made according to the formula

$$f_x = \frac{N_1 \cdot (f_0 \cdot k_m)}{N_2 \cdot k_m} \tag{7.50}$$

As $f_0 \cdot k_m = K = const$, the final equation will be:

$$f_x = \frac{N_1 \cdot K}{N_2 \cdot k_m} \tag{7.51}$$

In order to reduce the resulting calculation error, it is necessary to accept the following execution order for arithmetic operations of the same priority:

1) $N_1 \cdot K = k_1$

2) $N_2 \cdot k_m = k_2$

3) $f_x = k_1 / k_2$

The final number of bits for operands and results representation are determined by the range of converted frequencies $f_x \in [f_{x\,min}, f_{x\,max}]$, the range of numbers N_1, N_2 (Equations (7.48), (7.49) respectively), scaling and requirements of universality of arithmetic subroutines.

7.4.3   Error of T_{02} forming

The PCM analysis has revealed a number of additional components of the conversion error, connected with the forming of second reference time interval T_{02}. The error of forming the T_{02} includes two components: the error of the wavefront and the wavetail forming (Figure 7.16). In its turn, the last component includes the error due to the delay of reaction to interruption, and the error of the shift in time of the response for interruption.

The delay of the reaction to interruption (the time interval between the pulse of the frequency f_x on the interrupt input and the timer/counter start) is determined by the execution time for three instructions: (1) CALL to the interrupt vector; (2) the unconditional jump (JMP) to the interruption subroutine; (3) the start timer/counter:

$$\Delta\tau_{\text{delay}} = t_{\text{CALL}} + t_{\text{JMP}} + t_{\text{STRT}} \tag{7.52}$$

The delay $\Delta\tau_{\text{delay}} \geq 1/f_0$ is the reason for the systematic error because of the T_{02} reduction connected with the execution time of three commands mentioned above.

In its turn, the shift in time of the response for interruption also brings an additional component into the resulting conversion error distributed according to the uniform law of the distribution. The reason for this component is that the time used by the microcontroller for the response to interruption can vary. So, microcontrollers of MCS-51/52/196 families (Intel) answer to the interruption as a rule at the end of the current instruction cycle as well as for the majority of other microcontrollers. As the maximum instruction cycle, during which interruptions from f_x are possible (the program for forming the gate time T_{01}) is

$$\tau_{c_{\max}} = 2 \cdot \tau_{\text{cycle}}, \tag{7.53}$$

where τ_{cycle} is the machine cycle duration, and can vary, for example, in time reduction, so that the necessary time for the response to interruption is changed in limits $\Delta\tau'_{\text{delay}} \in]0; \tau_{c\max}]$.

The error of the T_{02} wavetail forming is determined by the delay $\Delta\tau''$, connected with the execution of the instruction of the logic polling for the last pulse of the frequency f_x. Its value is in the interval $\Delta\tau'' \in [3\tau_{\text{cycle}}, 5\tau_{\text{cycle}}]$ and is determined by the logic of the execution of the program part of the last pulse f_x polling. Then the

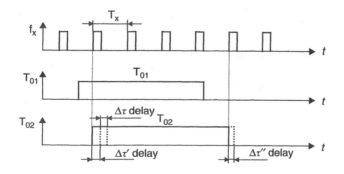

Figure 7.16   Reasons for instrumental error

real value of the gate time T_{02} is determined as

$$T_{02\text{real}} = T_{02} - \Delta\tau_{\text{delay}} - \Delta\tau'_{\text{delay}} + \Delta\tau''_{\text{delay}} \qquad (7.54)$$

The influence of the error of shift in time of the response for interruption and errors because of the T_{02} wavetail forming can be mutually compensated as the first component reduces the gate time T_{02} and the second increases it. The reduction of the error of shift in time of the response for the interruption can be achieved using a higher-speed microcontroller core or increased microcontroller clock frequency. In its turn, such measures will also result in the quantization error reduction as the frequency f_0 value in Equation (5.7) also will be increased.

7.5   Correction of PCM's Systematic Errors

The inclusion of the embedded microcontroller or the microcontroller core into smart sensors provides opportunities for the use of known methods of automatic error correction for measurement results, improvements to existing of methods and the creation of new methods.

The increase of the accuracy of frequency-to-code converters is possible due to the conversion algorithm improvement as well as to the use of additional data processing and calculation in order to correct conversion results. The first group of methods is focused on the reduction of methodical errors, the second group reduces both methodical and instrumentation errors. Thus PCM error reduction is possible at high conversion speed and without the essential algorithmic complication.

A number of additional error components revealed by PCM analysis can be reduced by systematic error elimination by a design stage of the PCM software.

As one of the main metrological features of frequency-to-code converters based on a PCM is the dependence of the conversion result on the PCM algorithmic structure and the software realization, then such methods increasing the conversion accuracy can be extended due to algorithmic methods of errors reduction alongside constructive-technological and protective-safety methods. The method, used for reduction of the gate time T_{01} forming error, because of the delay of reaction to interruption (Equation (7.52)) is one elementary way for the aprioristic elimination of systematic errors. The essence of this method consists of the change of the rigidly established instruction order for interrupts subroutine from external sources (converted frequencies f_x) by the service protocol for external interruptions. For the $\Delta\tau_{\text{delay}}$ reduction, the interruptions subroutine is modified so that whenever possible the instruction to the start timer/counter is executed as soon as possible.

For this aim, instead of the unconditional jump instruction, the start timer/counter instruction is placed in the memory cell of the external interrupt vector as specified by the service protocol of external interruptions. In its turn, the unconditional jump instruction JMP for the interruptions subroutine from pulses f_x is located in the next memory cell. Then $\Delta\tau_{\text{delay}}$ is determined by the execution time for only two instructions instead of three and is equal to

$$\Delta\tau_{\text{delay}} = t_{\text{CALL}} + t_{\text{STRT}} = 3 \cdot \tau_{\text{cycle}}, \qquad (7.55)$$

instead of $5\tau_{\text{cycle}}$.

As the systematic error $\Delta \tau_{\text{delay}}$ residue is no eliminated at this stage, and is still commensurable with the period T_0 of the reference frequency, it can be finally eliminated at the calculation stage of the converted frequency f_x according to Equation (5.4) with the help of the corrective action for N_2 by increment of its lowest-order byte.

Addresses of interrupt vectors in microcontroller families MCS-51/52 (Intel) are located in the resident memory with the interval in 8 bytes. Hence, this method of systematic error elimination can be successfully used for frequency-to-code converters based on this microcontroller family.

The instrumentation error $\Delta \tau'_{\text{delay}}$ can be reduced by constructive-technological methods because of the shift in time of the response for interruption: that is by using a higher-speed microcontroller or increasing clock frequency if possible.

The real value of the gate time T_{02} after elimination of the error component $\Delta \tau_{\text{delay}}$ will look like

$$T_{02\text{real}} = T_{02} - \Delta \tau'_{\text{delay}} + \Delta \tau''_{\text{delay}} \tag{7.56}$$

Thus, the method used for the algorithmic aprioristic elimination of systematic errors at in PCM software design reduces the error component $\Delta \tau_{\text{delay}}$ of the T_{02} wavefront forming practically without algorithmic and software complication. In its turn, further correction at the minimum algorithmic PCM complication (one additional one-byte instruction, INC ⟨register⟩) eliminates the time delay $\Delta \tau_{\text{delay}}$ completely.

As in this case, the condition for correction usage is determined by the inequality

$$\Delta t_{\text{correction}} \geq 1/f_0, \tag{7.57}$$

where $\Delta t_{\text{correction}}$ is the corrected time interval, the correction at the initial stage would not give an opportunity for full elimination of the error $\Delta \tau_{\text{delay}}$ because of the non-valid condition (Equation (7.57)). The combination of this method for aprioristic elimination of the error reasons and corrective action in the final result of the determination N_1 gives the effect of full elimination of $\Delta \tau_{\text{delay}}$. Thus, main systematic conversion errors can be eliminated by software design.

7.6 Modified Method of Algorithm Merging for PCMs

In designing multifunctional frequency (phase shift)-to-code converters for smart sensors based on modern microcontrollers, there is a problem of the resident memory economy, and consequently, the chip area. The task of memory size minimization for the PCM software consists of using methods of algorithm merging for union into a combined algorithms, reflecting the specificity of the solution in special situations. The precondition for this is a coincidence of some basic operations of elementary measuring procedures of PCM algorithms. However, existing methods of algorithm merging offered by Karp, Lazarev and Piyl' [141] as well as modification of Karp's algorithm [142], using partial matrix algorithm schemes, do not take into account, that during each combined algorithm execution, interruptions and the subroutine execution according to the required algorithms of functioning are possible. In the combined algorithm, it is also necessary to provide an opportunity of work with appropriate internal (from timer/counters) and external (hardware) interruptions at a limited number of interrupt vectors. Therefore, the method using partial matrix algorithm schemes was modified [143]. Thus, the modified method consists of the following:

1. According to initial block diagrams of two algorithms A and B, having common operators, partial matrixes $\|a_{ij}^*\|$ and $\|b_{ij}^*\|$, reflecting relationships of common operators are constructed. Their elements are represented by the Boolean function determining conditions of transition from the operator i to the operator j. The binary variable Σ is used for the indication of what should be realized from two variants of the combined algorithm. Let the value $\Sigma = 1$ correspond to the algorithm A, and the value $\overline{\Sigma} = 1$ to the algorithm B. The practical realization of the Boolean variable Σ is carried out simply enough by the set up or reset of the use bit.

2. Additional rows, N_{ij}, are included in partial matrixes of algorithms $\|a_{ij}^*\|$ and $\|b_{ij}^*\|$. Their elements are represented by Boolean variables determining the conditions of interrupts subroutines calls during the execution of appropriate operators of algorithms and reflecting relationships of common interruptions subroutines. If the interruption during the execution of the ith operator is enabled, the Boolean variable in this row is equal to '1', otherwise to '0'. Dimensions of partial matrixes of algorithms A and B are equal to $((q + N_{ij})N_A) \times ((q + N_{ij})N_B)$ accordingly, where q is the number of common operators; N_A and N_B are the number of all operators in algorithms A and B respectively.

3. In view of the additional row the partial matrix $\|m_{ij}^*\|$ of the combined algorithm with dimension $(q + N_{ij}) \times (N_A + N_B - q)$ is constructed

$$m_{ij}^* = \sum a_{ij}^* \vee \overline{\sum b_{ij}^*} \tag{7.58}$$

4. The partial matrix at the third stage is simplified whenever possible by extraction of the disjunctive equation

$$\left(\Sigma \vee \overline{\Sigma}\right), \text{ and if } m_{ij}^* = \left(\Sigma \vee \overline{\Sigma}\right) \text{ then } m_{ij}^* = 1 \tag{7.59}$$

Let's pass from the final partial matrix to the partial block diagram of the combined algorithm. Further, accepting that step by step $\sum = 1$ and $\overline{\sum} = 1$ and adding to the block diagram the rest parts of the block diagrams of algorithms A and B and we obtain the full block diagram of the combined algorithm. It is supplemented by conditional operators, whose dependence on the variable Σ enables or disables interruptions subroutines by the execution of this or that operator of the merged algorithm and modifies virtual interrupt vectors, thus providing the initialization of the required subroutine.

If united algorithms use different interruptions subroutines of a large size, it is expedient to merge algorithms of these subroutines with the help of Karp's modified method [142].

From the point of view of the theory of sets, the algorithms merging for PCMs of the frequency and the phase shift can be represented as the union of two sets, one of which is represented by the intersection of sets of operators of algorithms A and B, and the second by its symmetric difference:

$$(A \cap B) \cup (A \Delta B) \tag{7.60}$$

In its turn:

$$(A \Delta B) = (A \backslash B) \cup (B \backslash A) \tag{7.61}$$

where $A \backslash B = \{x : x \in A \text{ and } \overline{x} \in B\}$, $B \backslash A = \{x : x \in B \, \overline{x} \in A\}$.

Now is a good time to give a practical example. Let algorithm A be the algorithm of the optimum PCM, realizing the frequency-to-code conversion according to the ratiometric counting method. In its turn, algorithm B is the PCM algorithm for the phase shift-to-code conversion. The conversion method for the phase shift based on the determination of the average value t_x and the average value of the period T_x during the cycle T_{cycle}

$$T_{\text{cycle}} = n \cdot T_o + (0 \ldots T_x) = n \cdot T_x \tag{7.62}$$

is used as the PCM basis. Hence, the conversion time is fixed in the certain interval and multiplied by the period T_x. Due to this, the errors of the aliquant of T_{cycle} and the period T_x are eliminated. The frequency conversion range is extended down to infralow frequencies. Time diagrams of this method are shown in Figure 7.17.

Thus, there are the following features of using functional-logic possibilities of the microcontroller. The pulse counting of the reference frequency f_0 is carried out by the timer/counter. Pulses 1, 2 formed by the input shaper, come on the interrupt input INT. The reference time interval nT_0 is formed with the help of the program delay, and the interval T_{cycle} by means of logic analysis. Upon ending of the conversion cycle, we

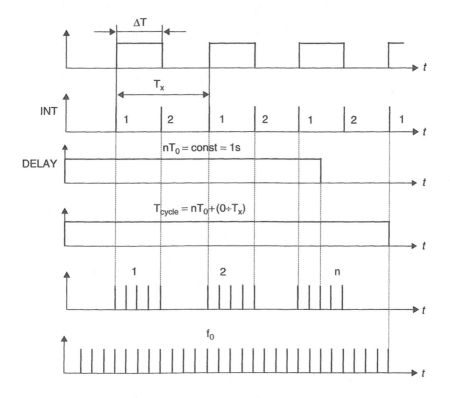

Figure 7.17   Time diagrams of phase-shift-to-code conversion method

receive two numbers

$$N_{tx} = \frac{t_x \cdot n}{T_0} \tag{7.63}$$

and

$$N_{Tx} = \frac{T_x \cdot n}{T_0} \tag{7.64}$$

The microcontroller calculates the average value of the phase shift according to the following equation:

$$N_{\varphi x} = 360 \cdot \frac{N_{tx}}{N_{Tx}} \tag{7.65}$$

The coincidence of basic elementary measuring procedures of algorithms A and B allows algorithm merging. Block diagrams of conversion algorithms for the frequency and phase shift are shown in Figure 7.18(a) and (b) respectively. Common operators are shown by identical symbols. Here $q = 7$, $n_A = 9$ and $n_B = 9$.

After execution of the first two stages of the algorithm merging, we obtain the following partial matrices:

$$\|a_{ij}^*\| = \begin{array}{c|ccccccccc} & F_0 & F_1 & F_2 & F_3 & F_7 & F_8 & F_9 & F_{10} & F_{12} \\ \hline F_0 & 0 & 1 & 0 & 0 & 0 & 0 & 0 & 0 & 0 \\ F_1 & 0 & 0 & 1 & 0 & 0 & 0 & 0 & 0 & 0 \\ F_2 & 0 & 0 & 0 & 1 & 0 & 0 & 0 & 0 & 0 \\ F_3 & 0 & 0 & 0 & \overline{P4} & P4 & 0 & 0 & 0 & 0 \\ F_7 & 0 & 0 & 0 & 0 & 0 & 1 & 0 & 0 & 0 \\ F_9 & 0 & 0 & 0 & 0 & 0 & 0 & 0 & 1 & 0 \\ F_{10} & 0 & 0 & 0 & 0 & 0 & 0 & 0 & 0 & 1 \\ N_1 & 0 & 1 & 0 & 0 & 0 & 0 & 0 & 0 & 0 \end{array} \tag{7.66}$$

$$\|b_{ij}^*\| = \begin{array}{c|ccccccccc} & F_0 & F_1 & F_2 & F_3 & F_5 & F_7 & F_9 & F_{10} & F_{11} \\ \hline F_0 & 0 & 1 & 0 & 0 & 0 & 0 & 0 & 0 & 0 \\ F_1 & 0 & 0 & 1 & 0 & 0 & 0 & 0 & 0 & 0 \\ F_2 & 0 & 0 & 0 & 1 & 0 & 0 & 0 & 0 & 0 \\ F_3 & 0 & 0 & 0 & \overline{P4} & P4 & 0 & 0 & 0 & 0 \\ F_7 & 0 & 0 & 0 & 0 & 0 & 0 & 1 & 0 & 0 \\ F_9 & 0 & 0 & 0 & 0 & 0 & 0 & 0 & 1 & 0 \\ F_{10} & 0 & 0 & 0 & 0 & 0 & 0 & 0 & 0 & 1 \\ N_1 & 0 & 1 & 0 & 0 & 0 & 0 & 0 & 0 & 0 \end{array} \tag{7.67}$$

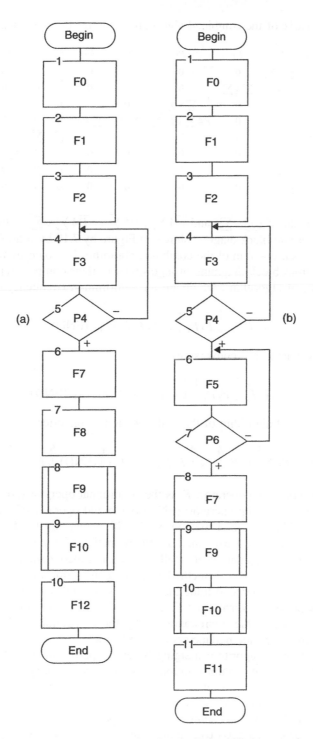

Figure 7.18 Block diagrams for frequency (a) and phase shift (b) to code conversion algorithms

The partial matrix of the combined algorithm is the result of the next stage:

$$
\|m_{ij}^*\| =
\begin{array}{c|ccccccccccc}
 & F_0 & F_1 & F_2 & F_3 & F_5 & F_7 & F_8 & F_9 & F_{10} & F_{11} & F_{12} \\
\hline
F_0 & 0 & \sum\vee\overline{\sum} & 0 & 0 & 0 & 0 & 0 & 0 & 0 & 0 & 0 \\
F_1 & 0 & 0 & \sum\vee\overline{\sum} & 0 & 0 & 0 & 0 & 0 & 0 & 0 & 0 \\
F_2 & 0 & 0 & 0 & \sum\vee\overline{\sum} & 0 & 0 & 0 & 0 & 0 & 0 & 0 \\
F_3 & 0 & 0 & 0 & \overline{P4}\sum\vee\overline{P4\sum} & \overline{\sum}P4 & \sum P4 & 0 & 0 & 0 & 0 & 0 \\
F_7 & 0 & 0 & 0 & 0 & 0 & 0 & \sum & \overline{\sum} & 0 & 0 & 0 \\
F_9 & 0 & 0 & 0 & 0 & 0 & 0 & 0 & 0 & \sum\vee\overline{\sum} & 0 & 0 \\
F_{10} & 0 & 0 & 0 & 0 & 0 & 0 & 0 & 0 & 0 & \sum & \sum \\
N_1 & 0 & \sum\vee\overline{\sum} & 0 & 0 & 0 & 0 & 0 & 0 & 0 & 0 & 0 \\
\end{array}
$$

After replacement of $(\sum\vee\overline{\sum})$ and $\overline{P4}\sum\vee\overline{P4\sum} = \overline{P4\sum}\vee\overline{\sum} = \overline{P4}$ it is possible to construct the partial block diagram shown in Figure by 7.19 by solid lines. The addition to the full block diagram of the combined algorithm is shown by the dashed line.

To move on from block diagrams of algorithms to their compact representation by the logic scheme of algorithms, we obtain for algorithm A (frequency conversion)

$$
F_0 F_1 F_2 \overset{1}{\downarrow} F_3 P_4 \overset{1}{\uparrow} F_7 F_8 F_9 F_{10} F_{12} \text{ (END)}, \tag{7.68}
$$

for algorithm B (phase shift conversion)

$$
F_0 F_1 F_2 \overset{1}{\downarrow} F_3 P_4 \overset{1}{\uparrow}\overset{2}{\downarrow} F_5 P_6 \overset{2}{\uparrow} F_7 F_9 F_{10} F_{11} \text{ (END)} \tag{7.69}
$$

and for the combined algorithm of the multifunctional converter

$$
F_0 F_1 F_2 \overset{1}{\downarrow} F_3 P_4 \overset{1}{\uparrow} P_5 \overset{2}{\uparrow}\overset{3}{\downarrow} F_7 P_7 \overset{4}{\uparrow} F_8 \overset{4}{\downarrow} F_9 F_{10} P_8 \overset{5}{\uparrow} F_{12} \overset{6}{\downarrow}\text{(END)} \overset{5}{\downarrow} F_{11} w \overset{6}{\uparrow}\overset{2}{\downarrow} F_5 P_6 \overset{3}{\uparrow} w \overset{2}{\uparrow}, \tag{7.70}
$$

where F_i is the processing operator; P_j is the conditional operator (conditional jump); w is the unconditional jump operator; (END) is the end operator of the logic scheme of the algorithm. This modified method of algorithm merging for PCMs has all the advantages of the modified Karp's method (the reduction of a number of operations at all stages due to the manipulation of smaller dimension matrixes), and also eliminates the inexactness inherent to Karp's method when a common operator in one algorithm is the last one, and in the other algorithm it is not. It also provides algorithmic functioning in conditions of possible interruptions. This method of algorithm merging reduces the resident memory size of the microcontroller by the realization of the multifunction converter for frequency-time parameters by two times.

Modified methods of algorithm merging based on partial matrix schemes of algorithms solves the problem of minimization for embedded memory in the smart sensor microcontroller or the microcontroller core only partially. By the synthesis of processor algorithms of the PCM there is the accompanying task of memory size minimization for the realization of algorithms composed from separate parts. With the aim to reduce the memory size, common parts can be allocated as subroutines $S_1, S_2, \ldots, S_\lambda$.

Let V_j be the memory size, occupied by the jth common part. For the subroutine initialization and return from it, it is necessary to execute two instructions: ACALL (or

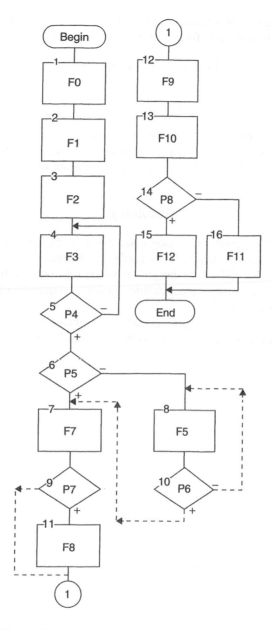

Figure 7.19 Block diagram for combined algorithm

LCALL) and RET. The memory size, occupied by the control instruction, is given by

$$W_j = V_{jCALL} + V_{jRET} \tag{7.71}$$

where V_{jCALL} is the part V_{lj}, related to each of programs using S_j (the call instruction for the subroutine CALL), and V_{jRET} is the part V_{lj} that belongs only to the subroutine S_j (the return instruction from the subroutine RET).

For real memory saving by subroutine allocation, the minimum number n_j of subroutine calls S_j from any programs from a higher level of hierarchy, is calculated as

$$n_j V_j > V_{lj} + V_{jCALL} n_j + V_{jRET} \tag{7.72}$$

From here

$$n_j > [(V_{lj} + V_{jRET})/(V_{lj} - V_{jCALL})] \tag{7.73}$$

If the jth part code of size V_j is common for the programs, the number of which is less than n_j, there will not be a gain in memory from the allocation in a subroutine and it is expedient to include their operators directly in the program body.

Summary

The use of PCMs for frequency-to-code conversion in digital output smart sensors with embedded microcontrollers reduces hardware, increases reliability, lowers the time-to-market period by the sensor design due to the reusing of software components and realizes sensors' self-adaptive capabilities (software-controlled performances and functional capabilities) in full.

In its turn, the method of algorithm merging reduces the memory size of the embedded microcontroller, which also contributes to hardware reduction.

8

MULTICHANNEL INTELLIGENT AND VIRTUAL SENSOR SYSTEMS

In most measuring and control systems, a large number of sensors is connected to a central computer or a microcontroller. The number of frequency–time-domain sensors is continuously increasing in different applications of sensor networks. However, the software interface from the frequency–time domain to the digital domain can be a complicated design problem.

8.1 One-Channel Sensor Interfacing

A sensor network with multilayer architecture, for example, as shown in Figure 8.1, is widely used in systems with distributed intelligence like a modern car. The advantage of such an architecture is that high layers use the information from lower layers and do not press in detail of the operation of lower layers [144,145]. This proposed architecture realizes some interfacing functions. First, it is a low-level hardware and software interface (sensor–microcontroller), secondly, it is a high-level controller area network (CAN) interface that has been developed for automotive applications to replace the complex cable in cars by a two-wire interface [144].

Some sensors with frequency output can be placed in a high impedance state when not required. This is useful for applications where input devices share a microcontroller. As a rule, outputs are microcontroller compatible and designed to drive a standard TTL or CMOS logic input over a short distance. If lines are greater than 30 cm, then it is recommended that a shielded cable is used between the microcontroller and the sensor. Therefore, the design of the hardware interface is solved relatively easily.

With software interfacing, it is necessary to ensure simple circuitry and design — standardization of the procedure of information interchange taking in to account the required speed. Parameters such as the quantization error and the processing time are in mutual inconsistency. There are the following methods of data exchange: program-controlled, with the use of interruptions (interruption on a vector) and direct access to memory.

The choice of software interface and the measurement technique in known traditional solutions depends on the desired resolution and the data-acquisition rate.

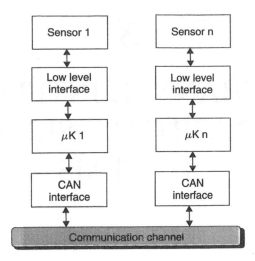

Figure 8.1 Multilayer sensor network architecture

For the maximum data-acquisition rate, period-measurement techniques are used. The maximum resolution and the accuracy are obtained using the frequency measurement. In comparison with known approaches, the program-oriented approach achieves universality. The output of the sensor is physically connected to the microcontroller input: the input of the timer/counter, the interrupt input or one of the lines of the input/output port. (The interface productivity depends on the selected type of connection.) The software interfacing module realizes the frequency (period)-to-code conversion at a virtual level inside the functional-logical architecture of the microcontroller. The conversion can be realized, for example, according to the method of the dependent count.

8.2 Multichannel Sensor Interfacing

Simultaneous multichannel measurements are necessary for connection of some frequency output sensors to the same microcontroller. The important feature of parallel information processing from several sensors is the multisequencing of the execution of elementary measuring procedures. This includes the detection of the internal parallelism as well as the conversion of sequential algorithms into quaziparallel, accepting the simultaneous execution of some elementary measuring procedures for implementation based on standard microcontrollers.

In planning parallel information processing in the microcontroller, it is necessary to take into account the time diagram. The rigid time diagram of operation superimposes limitations on the process of multisequencing of information processing and the choice of the functional-logical architecture of the microcontroller. Main indexes, on which the quality and efficiency of quaziparallel algorithms are evaluated, are the time and accuracy of measurement. These parameters are in mutual inconsistency. The method of the dependent count used for the frequency (period) measurement overcomes this inconsistency. As described above, except for high accuracy, the given method ensures the non-redundant time of the frequency-to-code conversion.

Further description of multichannel sensor interfacing involves an example of an automobile sensor network, consisting of four sensors of rotation speed with frequency output. This application is relevant for the automotive antilock braking system (ABS). It has enhanced vehicle safety on slippery roads [146]. Braking on surfaces with poor road adhesion is less effective, therefore excessive braking pressure can cause the wheels to lock. However, this effect leads to serious diminishment of adhesion coupled with the loss of lateral grip. To avoid this, an electronic system measures the wheel speeds using rotary speed sensors, and by means of solenoid valves ensures that the braking pressure is briefly reduced when a value of zero is measured. Based on the simultaneously registered position of the brake pedal, the intention of the driver can be determined and the brake pressure built up again accordingly. Where necessary, this process is repeated. While the first antilock brakes introduced at the end of the 1970s resembled automatically actuating cadence braking systems, today they are capable of regulating the wheel with a high degree of precision in a range just prior to the onset of locking.

A few years ago cars had become computer networks on wheels. The average car had about $653 worth of electronics and/or electromechanical systems, about 15% of the total value of the car. Now the average car contains about $950 to $2000 worth of electronics, depending on whether the valuation is based on the cost or the price. It is estimated that electronics makes up 25 to 30% of a car's total value [147]. Automotive microsystem applications, especially for antilock braking systems, require highly reliable parts that can withstand a harsh environment and wide temperature range. On the other hand, their price should not be excessively high in conditions of large production volumes.

The reliability of the ABS is determined by the speed of processing of the measuring information and the solution made by the control system. In other words, the ABS must work in real time. However, in many ABS, the measurement of the rotation speed is based on conventional methods of the frequency measurement (standard direct and indirect counting methods) or on the reciprocal (ratiometric) counting method.

In order to design a reliable and low-cost ABS it is necessary to take into account three main components of the ABS: the rotation speed sensor, the encoder and the method of measurement (Figure 8.2) [148]. Let us consider all these components step by step.

8.2.1 Smart rotation speed sensor

Sensor technology is playing a critical role in the development of new products and can govern the feasibility of deploying certain systems [149]. Critical sensors involved are those for sensing the wheel rotation speed [150]. Three types of rotation speed sensors can be used in the ABS: Hall sensors; the active semiconductor sensor (both are modulating sensors) and the passive self-generating inductive sensor.

The active semiconductor sensor with frequency output has been selected as the wheel rotation speed sensor [52]. The rectangular impulse sequence is generated as the sensor's output.

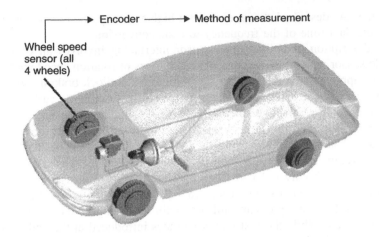

Figure 8.2 Three main components of ABS wheel speed sensing concept

As the rotation speed is connected with the sensor's output frequency by the following dependence:

$$n_x = f_x \cdot \frac{60}{z} \tag{8.1}$$

(where Z is the number of teeth), its amplitude is constant and does not depend on temperature and direction of the wheel rotation. Such a signal can be used at zero speed, without any prior processing at the input (filtration, digitization, etc.). The maximum amplitude of the output signal is determined by the voltage supply (in the case of a sensor without an input block) or by the supply voltage of this input block. The sensor is not influenced by run-out and external magnetic fields in comparison with Hall sensors. With Hall sensors, it is also necessary to take into account the availability of the initial level of the output signal between electrodes of the Hall's element by absence of the magnetic field and its drift. It is especially characteristic for a broad temperature range, defined by the road climatic conditions of various countries.

The main performances of the active semiconductor sensor are:

- the 50% duty cycle
- the non-contact operation in a heavy duty environment
- no magnet
- any metallic target: steel, copper, brass, aluminium, nickel and iron.

The active inductive microsensor [55] is another suitable sensor. The sensor offers a combination of interesting features, such as position, speed and direction information. It is composed of two silicon chips, one for the integrated microcoil and the other for the integrated interface circuit.

Oscillograms of the sensor's output signals are shown in Figures 8.3–8.5. It corresponds to the sensor operation for 200, 1500 and 3000 rpm accordingly. The supply voltage for the sensor was +12 V. The sensor has demonstrated the steady operation

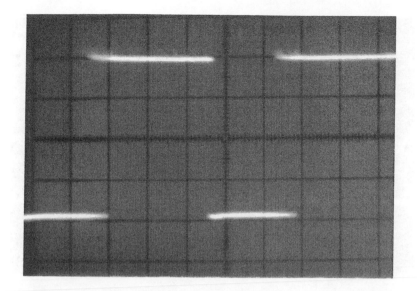

Figure 8.3 Oscillogram of sensor's output at ~200 rpm (1 V/div and 2 ms/div)

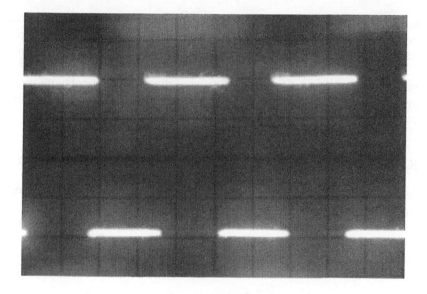

Figure 8.4 Oscillogram of sensor's output at ~1500 rpm (1 V/div and 0.2 ms/div)

during change of the supply voltage from +5.7 up to +30 V. All measurements were carried out with a working air gap of 1.5 mm.

8.2.2 Encoder

The use of Hall sensors and special encoders in ABS increases the cost of the system, especially for four wheels because such encoders consist of polarized magnet parts. Due

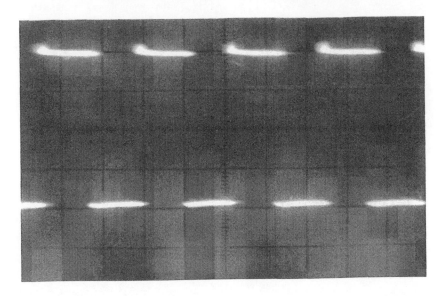

Figure 8.5 Oscillogram of sensor's output at ~3000 rpm (1 V/div and 0.2 ms/div)

to the use of the active semiconductor wheel speed inductive sensor, the encoder for the proposed sensor microsystem does not require complex construction with oriented magnetic particles. The encoder can be made from cheaper material, for example, soft steel.

Geometrical teeth sizes of the encoder and one of the possible orientations are shown in Figure 8.6 and mild steel encoders in Figure 8.7. Some variations of geometrical teeth sizes and the orientation are possible.

8.2.3 Self-adaptive method for rotation speed measurements

The described ABS is based on the method of the dependent count (the method with a non-redundant reference time interval) in which simultaneous synchronous-cyclic frequency measurement of sensor signals is carried out. This frequency is proportional to the rotation speed of the wheels. The measurement time for this method is the minimum possible. The quantization error as well as the measurement time practically

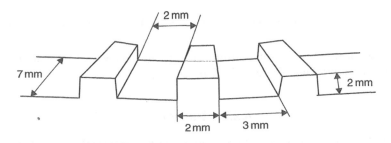

Figure 8.6 Teeth geometry for still encoder

Figure 8.7 Mild steel encoders

depend neither on rotation speed nor exceed the above given values. Therefore, the sampling rate of the output signal from the sensor ensures the arrival of information in all four channels in real time.

Let us determine this sampling rate. Let the automobile drive at a speed of 240 km/h, i.e. 67 m/s, and the time of measurement should not exceed 0.1 s (i.e. for this time the automobile will drive no more than 8.7 m at maximum speed). Let us define the quantization time having set the relative error of measurement from 0.05% to 0.5% and the reference frequency $f_0 = 1/T_0 = 1$ MHz:

$$T = (1\text{--}2) \cdot T_0 N = (1 \ldots 2) \frac{T_0}{\delta}, \tag{8.2}$$

where N is the minimum number of impulses, which should pass to the reference frequency counter according to the necessary relative error δ. This time interval will vary within the limits 2–4 ms at $\delta = 0.05\%$ and 0.2–0.4 ms at $\delta = 0.5\%$. When $f_0 = 10$ MHz this time can be reduced 10 times. Therefore, during 0.1 s 25–50 measurements per second at $\delta = 0.05\%$ and 250–500 at $\delta = 0.5\%$ will be carried out.

Therefore, the four-channel frequency-to-code conversion of the ABS control block will execute simultaneous synchronous frequency measurements in real time. The high sampling rate of the signal reduces the specific measurement time to 0.1 s without decrease in accuracy.

Other possibilities, e.g. the automatic choice of the reference time interval depending on the given error of measurement, require an advanced ABS algorithm. Whereas speed is one of the major ABS performances, the required error of measurement can be selected by the microcontroller depending on the current rotation speed due to the adaptive possibilities of the method. It increases the speed by measuring of critical rotation speeds.

The measurement system will also function successfully in the absence of wheel slip and the rotation speed can vary in a wide range: from zero speed up to a maximum. The essential advantage of this method is the possibility of digital measurement of acceleration with similar high accuracy and without extra circuitry. It opens the possibility of using the control method not only for speed, but also for acceleration of rotation as well as the combined control method. There are opportunities to develop the modern high-end ABS for future needs. For example, road hazards may cause the ABS to function unexpectedly. Using the proposed approach, ABS will compensate for road conditions or poor judgement.

Due to the minimal possible circuitry necessary for the sensor output processing, the metering microcontroller core can be easily embedded into the microsystem. The next step of the microsystem creation is integration of the microcontroller core with necessary peripherals into the system design. The most preferable microcontrollers for connection of four rotating speed sensors with frequency outputs are microcontrollers that have four or more external interrupts and two timers/counters. Microcontrollers of the MSP430 family (Texas Instruments) and some microcontrollers of the MCS-51 family (Intel), for example, 8XC51 GB are suitable.

8.2.4   Sensor interfacing

Four measuring channels are subdivided conditionally into two group pairwise channels. This enables rational planning of multisequencing of the algorithm for multichannel measurement and effective utilization of features of the microcontroller's functional-logical architecture. Each of the two groups functions identically. Time diagrams of the parallel measurement of frequencies of two sensors from one of the groups are shown in Figure 8.8.

Let's consider the operation of one of the groups in more detail. Let the frequency be $f_{x1} > f_{x2}$. The multichannel method of measurement is realized in the following way.

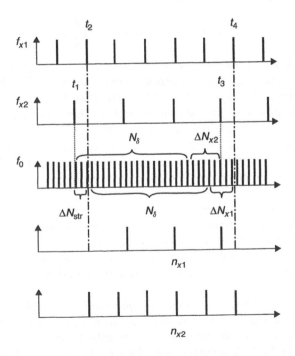

Figure 8.8   Time diagram of two-channel simultaneous measurement

Impulses of both frequencies are calculated by interrupts. The internal clock reference frequency ($f_0 = F_{osc}/12$, where F_{osc} is the clock frequency of the microcontroller) is calculated with the help of the built-in timer/counter. The number N_δ is written beforehand into the register of the timer/counter. It is determined by the required accuracy of the measurement $N_\delta = 1/\delta$, where δ is the relative quantization error. With the arrival of an impulse of the smaller frequency (time moment t_1) the timer/counter begins to form the reference time interval, which is determined by the number N_δ. Impulses of frequency f_{x2} are calculated in parallel. At the time moment, when the forming of the reference time interval determined by the number N_δ has completed, at the moment of the arrival of the following impulse of frequency f_{x2} the count of impulses of sequences with frequencies f_{x2} and f_0 in the first channel finishes (the time interval t_3). The number ΔNx is fixed in the timer/counter after its overflow or in one of the microcontroller's registers. In parallel to the forming of the time diagram of measurement for the first channel, the time diagram for the second channel is formed.

The measurement in the second channel begins with the arrival of the impulse of the greater frequency f_{x2} (time moment t_2). However, in this moment the timer/counter has already formed part of the reference time interval (between the moment t_1 and t_2), determined by the number ΔN_{str}. This number is read out from the timer's register and is stored at the beginning of measurements (the time moment t_2) in the second channel with the purpose of consequent correction. After counting the specific number N_δ in the first channel the number ΔN_{str} will be written into the timer/counter allowing the same number to be counted in the second channel. Then up to the time moment t_3 the number $\Delta N_{x1} - \Delta N_{str}$ is calculated, and up to the moment t_4 the number ΔN_{x2} is calculated. The result of the measurement in each of the channels is calculated from:

$$f_{x1} = \frac{n_{x1}}{N_\delta + \Delta N_{x1}} \cdot f_0 \tag{8.3}$$

$$f_{x2} = \frac{n_{x2}}{N_\delta + \Delta N_{x2}} \cdot f_0, \tag{8.4}$$

where the n_{x1}, n_{x2} are the integer numbers of impulses with frequencies f_{x1}, f_{x2} calculated by the interrupts.

The measurement time for each of the channels is always equal to an integer of the periods of measurand signals:

$$t_{x1} = \frac{n_{x1}}{f_{x1}} = \frac{N_\delta + \Delta N_{x1}}{f_0} \tag{8.5}$$

$$t_{x2} = \frac{n_{x2}}{f_{x2}} = \frac{N_\delta + \Delta N_{x2}}{f_0} \tag{8.6}$$

Relative quantization errors for each of the channels are calculated as:

$$\delta_1 = \frac{1}{N_\delta + \Delta N_{x1}} \tag{8.7}$$

$$\delta_2 = \frac{1}{N_\delta + \Delta N_{x2}} \tag{8.8}$$

These errors do not exceed the specified relative error δ. Thus, the hardware multi-channel sensor interfacing will be realized with the help of four external interrupts and two built-in timers/counters. The low-level software interface and simultaneous measurement with the measurement non-redundant time and the constant error in all specified measuring ranges of frequencies will be realized with the help of the program-oriented method of measurement. The measuring channel will be realized an the virtual level inside the functional-logical architecture of the microcontroller.

This solution of sensor interfacing can be used in the high-end (Figure 8.9) and the mid-range (Figure 8.10) antilock braking system in an automotive application. The centralized architecture with one microcontroller can be used for the low-cost (four wheels) mid-range ABS. The multilayer architecture is suitable for the high-end ABS with four microcontrollers and one computer. In this case, the sensor element, the embedded microcontroller and the circuitry for the CAN interface are realized as a sensor microsystem (one for each wheel).

Computer

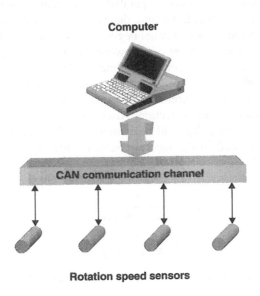

CAN communication channel

Rotation speed sensors

Figure 8.9   Multisensor architecture for high-end ABS

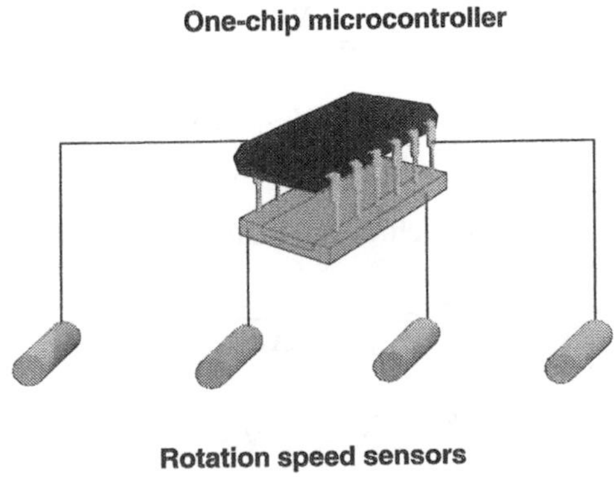

One-chip microcontroller

Rotation speed sensors

Figure 8.10   Multisensor architecture for mid-range ABS

The following advantages of the rotation speed measurement concept are achieved:

- cost reduction

- more simple and cheaper encoders

- possible to be used in all modern wheel and sensor assemblies and realized in many variations

- economic performance (greater efficiency in the ratio productivity/manu-facturability/cost)

- high metrological reliability by miniaturization, weight saving and zero speed capability, and, consequently, increased car safety

- possible to use the advanced ABS algorithm.

8.3   Multichannel Adaptive Sensor System with Space-Division Channelling

The automatization of complex processes and scientific research requires the development of multichannel reliable sensor systems with high performances and functional capabilities, working in real time. It is especially important when such systems are unique sources of objective and complete information. The main requirements for modern sensor systems are the following:

1. High software-controlled accuracy, for industrial and scientific (precise) measurements, and its constant values in a wide dynamic frequency range.

2. The non-redundant time for frequency-to-code conversion by the software-given relative error of conversion.

3. The multichannellness of systems with time and, especially, with space separation of channels for simultaneous conversion in all channels.

4. High speed and noise immunity for transmissions of signals.

5. High reliability, including metrological, by simple circuit realization.

6. Total automatization and various kinds of adaptation.

7. Possibilities for functional and structural tuning.

8. Self-checking and self-testing.

To achieve all the above listed requirements in one sensor system is very difficult, and results in a high degree of complexity. Also many requirements are inconsistent, and sometimes, mutually exclusive.

A possible smart sensor system can be based on centralized and decentralized principles, with the use of local information-controlled subsystems, software-controlled number of channels, regimes of data transmission (cyclic, sporadic, on call); with backward channels or without etc. The frequency–time-domain signal is used as the informative signal parameter.

The method of the dependent count for the frequency (period)-to-code conversion with software and the hardware–software realization can be successfully used in such a sensor system. For transmission of digital signals, the method of the digit-by-digit phase manipulation can be used. The sensor system consists of two main modules: the software-controlled multichannel module for data acquisition and the processing of primary information with the space and/or time separation of frequency channels; and the module for transmission of digital signals.

The first module is based on the non-traditional method of simultaneous acquisition and processing of frequency measuring signals for n channels. This method makes available the execution of the physical and algorithmic measuring procedures with software-controlled accuracy. It provides fast processing in a wide dynamic range from low to high frequencies, exceeding the reference frequency. The speed is the greatest possible because the time for conversion and the digital processing are non-redundant for the whole frequency range. It is determined by the given accuracy and is automatically established during the conversion, and does not depend on frequencies and numbers of channels. The proposed method differs from conventional DSP methods, where it is necessary to complete a large number of calculations in real time.

One possible way of realization of the new method for multichannel simultaneous processing of frequency signals is shown in Figure 8.11. The system contains a microcontroller, n blocks for the Dirichlet window forming according to the method of the dependent count (BFDW), the digital multiplexer (MX), channel memory registers (MR) with strobing inputs Ci, the synchronization block of conversion and processing (SCP), the logic element OR and the binary encoder (CD) for channel numbers in order to control the MX [151].

Modules BFDW are identical for any n channels. The BFDW is shown in Figure 8.12. It consists of two logic elements AND, two D-triggers and the counter CT. Input pulses f_1 through the AND$_1$ arrive on the decrement count input of the CT.

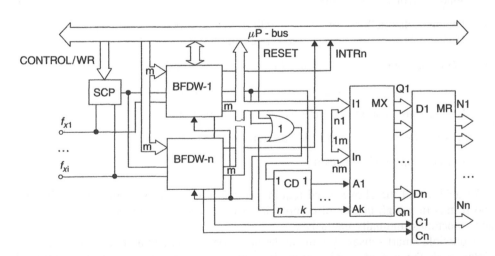

Figure 8.11   Circuit diagram of sensor system with space-division channelling

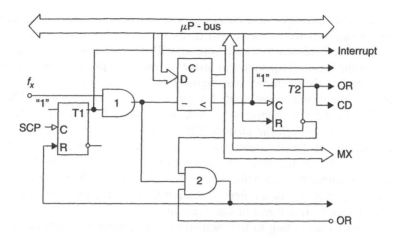

Figure 8.12 Circuit diagram of block for dirichlet window forming

The counter capacity m must be calculated according to the following formula:

$$m = \text{ent}\left[\lg_2\left(k \cdot n \cdot N_i \cdot \frac{f_{max}}{f_0}\right)\right], \tag{8.9}$$

where $k = 2$–3 is the factor determined by those tasks, which are executed by the microcontroller by the processing of the multichannel frequency signal (numerical processing of results of previous conversions, functional control, etc.); N_i is the number of clock cycles for the interrupt subroutine execution according to the interrupt vector; f_{max} is the maximum possible frequency in any from n channels. It can be more or less than f_0.

During the pulse count, the counter CT is periodically overflowed, which causes the interrupt. The interrupt routine executes the dynamic frequency-to-code conversion and requires N_i clock cycles, according to Equation (8.9). Before the start of the frequency-to-code conversion with the given relative error δ, the microcontroller calculates the number $N = 1/\delta$, which can be decomposed into the linear combination:

$$N = N_a + N_i \cdot 2^m, \tag{8.10}$$

where m is the counter capacity. N_a are recorded in the counter for each channel, and N_i are stored in the microcontroller's memory. The input frequency signal f_{xi}, signals CONTROL and WRITE enter the block SCP.

According to results of the previous processing, the SCP determines the minimum input frequency, on which the measuring conversion will begin. Signals INT$_i$, which will hinder the triggers turn on to '1' by the counter reset are simultaneously set up. After the signal RESET elements AND are opened through the element OR and triggers in each channel are set up into '0' by the first pulses of input frequencies. After this preparation, the measuring conversion according to the method of the dependent count will be started immediately. The microcontroller forms the signal WRITE and the block SCP forms the signal, on which triggers are switched on. Elements AND are thus opened and pulses of frequencies f_{xi} enter the counters. At each reset, interrupts INT$_i$, are formed and the timer/counter decrements the numbers N_i that were recorded into the microcontroller beforehand and checks the equality to zero. So, the decomposition of the pulse count into the software and hardware in the ith counter is realized. It is obvious, that the hardware-software counter will be reset first in that channel, the input of which the higher frequency will arrive (it has taken place in the first channel). The microcontroller then removes the signal INT_i on the D-triggers T_2 input and consequently at the next counter reset this trigger will be set up to '1'. The logical element AND_2 in the ith channel is closed, and similar logical elements of other channels are opened through the element OR. Simultaneously the encoder forms the code of the address on the multiplexer's inputs so, that the digital code from the Q-output of the counter of the zeroized channel arrives on D-inputs of registers MR. As far as the assumption made before the logical element AND_2 is closed, the pulse count of the frequency f_{x1} in the first channel will be continued up to the ending of measuring conversions in the rest of the channels.

In these channels according to the method of the dependent count, the first pulse of the input frequency, which feeds in the counter after the reset in the first channel, through the element AND_1 resets to the initial condition the D-trigger T_1 and the frequency-to-code conversion in this channel is stopped up to the next measuring cycle. Thus, the pulse will be coming on the strobe input of the ith register. This pulse rewrites the current code into the register from Q-outputs of the counter CT, as well as the request on interruption from the output of the trigger T_1, which testifies the ending of the conversion in the ith channel. Thus, after ending the pulse count in all channels, except the first one, the microcontroller generates the signal RESET and the signal INT_1 again, which sets up the trigger T_2 to '0'. The signal RESET through the element OR and the opened element AND_2 reset the trigger T_1 to '0'. Then, the element AND_1 will be closed and the pulse count of the frequency f_1 is complete.

Codes of remainders from subtraction of the number of pulses N_i of the input frequency of the ith channel from the initial number $N = 1/\delta$, which was written before will be in hardware-software counters:

$$N_i = N - N_i', \tag{8.11}$$

where N_i' is the number of pulses in the counters of the ith channel at the moment of measuring conversion ending. In its turn, in the ith register, the codes R_i, will be written. These are remainders from subtraction of the pulse number R_i of the maximal frequency from the initial number N during the time interval forming in the BFDW of the ith channels, i.e.

$$R_i = N - R_i', \tag{8.12}$$

Pulses generated by BFDW guarantee the integer number N_i of periods T_{xi} of the input frequency $f_{xi} = 1/T_{xi}$ in the ith channel. Due to this, the time interval T is multiple to the period T_{x1}. The methodical quantization error is excluded in all channels with frequencies less than f_{x1}. Therefore, the following equation is true with the absolute maximum quantization error Δ_q:

$$N_i/f_i = R_1/f_1 \tag{8.13}$$

The relative quantization error is the same as in the one-channel frequency-to-code converter, according to the method of the dependent count $\delta_q < \delta = 1/N$.

The cycle duration for the signal processing of all frequency signals is determined by the duration of conversion for the maximum frequency to code, equals $T_1(N + \Delta N)$ and the processing of $T_{processing}$ for measuring codes of all channels.

Such a method for multichannel acquisition and processing of frequency signals simultaneously in all channels opens the way for the design of new simple and highly reliable multichannel sensor systems for precise measurements with software-controlled performances and functional facilities. On the basis, it is possible to create multichannel digital sensors for different physical and chemical values and consequently, sensor networks for transmission of the digital information.

8.4 Multichannel Sensor Systems with Time-Division Channelling

One possible variant of the multichannel data acquisition system with time-division channelling for temperature frequency output sensors is shown in Figure 8.13.

The converter's structure depends on the conversion method used, which should provide high metrological characteristics. In this device the method of the dependent count, which provides the software-controlled polling alongside the time-shared cyclic, was used. Thus, the time of measurement $T_{measurement}$ does not exceed the value

$$T_{measurement} = N_q \cdot (1 + \Delta k)T_0 \qquad (8.14)$$

where N_q is the parameter determined by the given quantization error $\delta_q = 1/N_q$; Δk is the variable, with value in limits $0 <= \Delta k < 1$.

The device works as follows. At the signal 'Start' the microcontroller chooses the number of the channel N_{ki} of the converted frequency f_{xi} and determines the parameter N_{iq} according to the given quantization error. At signals 'Setting Up' (a, b) triggers T_2, T_3 and the counter CT_2 are nulled. The complement of the number N_{iq} is put into the counter CT_1. The first pulse of the frequency f_{xi} on the multiplexer's MX output, sets by its wavetail the trigger T_2 to '1' through the logical element AND_2. The enable signal from the trigger opens elements AND_4 and AND_5, and counters CT_1 and CT_2 start the pulse counting of the reference f_0 and converted frequencies accordingly. When the number of pulses of the frequency f_0 is compared to the number N_{iq}, the signal from the counter's output CT_1 sets up the trigger T_3 to the '1'. The logical '1' from the trigger's output opens the logical element AND_3, and the wavetail of the next pulse of frequency resets the trigger T_2. Thus, numbers N'_{f0i} and N_{fxi} are accumulated in counters CT_1 and CT_2 accordingly. Having ended the processing of results of measurement in the $(i - 1)^{th}$ channel, the microcontroller repeats the described cycle of measurements in the $(i + 1)^{th}$ channel.

The result of the previous measurement, carried out in parallel with conversion, consists in the temperature determination according to the frequency f_{xi} for the ith temperature-to-frequency converter and correction of results. Thus the following equations are used:

$$N_{f0i} = N'_{f0i} + N_{iq}; \qquad (8.15)$$

$$f_{xi} = \frac{f_0 \cdot N_{fxi}}{N_{f0i}}; \qquad (8.16)$$

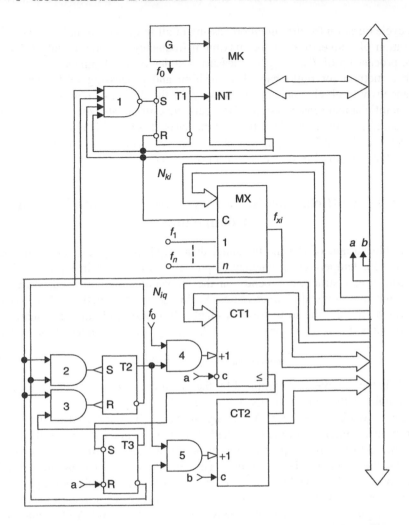

Figure 8.13   Multichannel sensor systems with time-division channelling

$$t^0C = K_i \cdot f_{xi}, \tag{8.17}$$

where N_{f0i} is the number of pulses of the reference frequency; K_i is the factor determined as the result of the correction subroutine execution. Its value depends on the sensor type, the measuring converter, the method of correction and the part of a sensor characteristic within the limits of which there is the required value of temperature.

The industrial data acquisition system based on the described principle is six-channel and intended for temperature measurement in the range $0–500\,^\circ$C in zones of material and press form heating of the moulding press in the technological process of plastic production. The relative error of the frequency measurement by the reference frequency $f_0 = 1$ MHz and $f_{x\,max} = 6$ kHz is equal to 0.05% in all dynamic ranges. This error can be reduced if necessary.

8.5 Multiparameters Sensors

Modern microelectromechanical and semiconductor technologies enable the integration of microelectronics circuits and multifunctional sensor arrays fabricated on silicon substrates for detecting different kinds of chemical and physical parameters. Such sensors are called *multiparameters sensors* and represent a single construction, and detect a set of physical or chemical quantities, concentrated in a small, local area. It is not a simple unit of one-functional sensors, but rather its structurally advanced combination is created with the aim to reduce the chip area and share the use of the digital or quasi-digital output, etc. Advanced sensors for detecting simultaneously various parameters such as temperature, pressure, gas and vapour concentration, odour, acceleration, inertia, electric and magnetic fields, etc. can generate frequency–time-domain output signals carrying the information provided by sensing elements. Modern silicon technologies offer many advantages for the design of multifunctional (multiparameters) smart sensors. It is expedient to have multisensor arrays in electronic noses and tongues as well as in medical implemented sensors; for temperature, pressure and humidity detection in one multiparameters sensor for different environmental tests and a very broad range of applications (e.g. biotechnological, food industry, etc.).

If the frequency in some channels has linear functional dependence on several measurands such parameters influencing the multiparameter frequency sensors are calculated by special algorithms, for example, described in [152]. For example, in the case of two-parameters frequency sensor, there are the following straight-line conversion characteristics:

$$\begin{cases} F_{x1} = a_1 y + b_1 x \\ F_{x2} = a_2 y + b_2 x, \end{cases} \tag{8.18}$$

where a_1, a_2, b_1, b_2 are the factors of sensitivity; x, y are detecting parameters. For the determination of x and y the following calculation algorithm should be used:

$$\begin{aligned} y &= T_1 \cdot F_1 - T_2 \cdot F_2 = N_1 - N_2 \\ x &= -T_3 \cdot F_1 + T_4 \cdot F_2 = -N_3 + N_4, \end{aligned} \tag{8.19}$$

where

$$T_1 = \frac{b_2}{\Delta}; T_2 = \frac{b_1}{\Delta}; T_3 = \frac{a_2}{\Delta}; T_4 = \frac{a_1}{\Delta}; \Delta = a_1 \cdot b_2 - a_2 \cdot b_1$$

This calculation can be realized by any microcontroller or the DPS microprocessor.

In general, signal processing and conversion for multiparameters sensor signals are similar to the processing used in multisensor systems.

One example of multiparameters sensors is the MPS-D sensor from SEBA Hydrometrie GmbH for the simultaneous measurement of six parameters: pH value, redox potential, conductivity, temperature, water level and dissolved oxygen [153]. The sensor has a digital output for connection to the data logger MDS (RS-232), directly to the PC, or to data transmission units (modem, radio-modem, etc.).

8.6 Virtual Instrumentation for Smart Sensors

The processing and interpretation of information arriving from the outside is one of the main tasks of data acquisition (DAQ) systems and measuring instruments on a

PC basis. As a result, the problem of adequate interfacing of various frequency–time-domain (quasi-digital) sensors with the PC for data acquisition arises for developers and users of any DAQ system [154].

This essentially facilitates the task of the interface design, which should be able to mimic traditional instruments to provide easy-to-use conventional understanding for non-experts [155]. That is, all the beneficial features of a PC, such as the programmability, the digital processing and storage capabilities are accessible to the user. A solution of these problems can be found by utilizing a personal computer to control smart sensors via special hardware and by interfacing with the user via a custom-built graphical user interface. The data acquisition hardware must have many digital input lines, entirely reconfigurable under the software control. As this system was developed, it effectively became a virtual instrument dedicated to the development of smart sensor systems.

It is known that the 'DAQ hardware without the software is useless — and the DAQ hardware with the poor software is almost as useless' [7]. It is especially true for today, when the obvious lag of algorithms and the software development from the progress in microelectronics is observed.

There has been a great deal of progress during the last decade in the development of virtual measurement technology. As a result, nowadays we have the so-called virtual instrumentation revolution [156]. Appreciable progress is also observed in the area of microcontroller-based intelligent measuring instruments. Such instruments have graphical displays and embedded microcontrollers 386EX (Intel). This microcontroller permits computing capabilities comparable with the power of the general-purpose PC computer. Printing and storing data via the RS-232/485 or IEEE-488 interfaces and the software to download the acquired data and screens from or to an MS-DOS or Windows compatible computer for further processing permit such instruments to be easily integrated in computer-based measuring systems. Thus, the technological progress in microelectronics and the specialized software environment means that the distinction between the virtual instrument, intelligent measuring instruments and measuring systems with the graphical user interface (GUI) becomes slightly fuzzy.

Though virtual instruments (VI) have been used in modern instrumentation since the mid-1980's, nevertheless, the generally accepted terminology related to the virtual instrumentation does not exist as noted in [157]. Two manufacturers, Hewlett-Packard and National Instruments form the USA, introduced virtual instruments for the first time into the market. There are two definitions of virtual instruments proposed by them.

According to National Instruments [7,156], 'Virtual Instrument — (1) A combination of hardware and/or software elements, typically used with a PC, that has the functionality of a classic stand-alone instrument; (2) A LabVIEW software module (VI), which consists of a front panel user interface and a block diagram program'.

According to Hewlett-Packard [158], the capability of using of graphical software and a personal computer for the processing and displaying of measurement results has been referred to as 'virtual instrumentation'. This term can be used to describe the following four areas: an instrument system as a VI; software graphical panels as a VI; graphical programming technique as VI; reconfigurable building block as VI.

The existing definitions of a virtual instrument, proposed by National Instruments and Hewlett-Packard are rather broad and include measuring systems with GUI. Other definitions, for example [159], cover only internal plug-in data acquisition boards. Besides, the part (2) of the National Instruments' definition includes only the LabVIEW software module. However, virtual instruments can be successfully realized with the help of ComponentWorks or LabWindows/CVI software from National Instruments. The strictest and the most successful virtual instrument definition was formulated in [157]: 'The Virtual Instrument can be described as an instrument composed of a general-purpose computer equipped with cost-effective measurement hardware blocks (internal and/or external) and software, that perform functions of a traditional instrument determined both by hardware and software, and operated by means of specialized graphics on the computer screen'.

The author of this definition proposes to interpret this definition in a strict sense. It means that the equipment can be qualified as a virtual instrument, when its hardware part and software part cannot operate separately as a measuring instrument. Then the software creates a new metrological quality — creates a new measuring function. Such an interpretation enables one to differentiate a virtual instrument from a measuring system with a GUI.

However, in our opinion, it has one defect, i.e. the unwieldiness. Such a definition as a distinction from that offered early, differentiates a virtual instrument from a measuring system with a GUI, because it means that the hardware part and software part cannot operate separately as a measuring instrument. In our opinion, the last phrase is the key. However, it has not been included in the definition, though it is met in the text repeatedly [157]. Besides, today's measuring science includes such new conceptions as the 'virtual measuring channel' and 'program-oriented methods of measurement'.

With allowance, the proposed 'mathematical' definition of a virtual instrument can be formulated as [160]:

> 'The *necessary condition* of the VI existing is the software realization of the user interface, performed by the general-purpose computer and the *sufficient condition* is that the hardware and the software part of the VI do not exist separately as an instrument.'

According to [155] 'in the present context 'virtual instruments for intelligent sensors' imply the utilization of microcomputers with the special hardware and software in development, in production and in field control, test and calibration of an intelligent sensor'.

8.6.1   Set of the basic models for measuring instruments

It is known that a typical measurement channel includes three main parts:

- input data acquisition
- data processing
- output data presentation (user interface).

As shown in [157] methods of each function of realization of the measuring channel permit us to classify a measuring instrument as a traditional, virtual or measuring system with a GUI. Three types of the realization can be described as the following:

$$R_1 = F_1(H) + F_2(H) + F_3(H) \Rightarrow TI \tag{8.20}$$

$$R_1 = F_1(H) + F_2(H) + F_3(S) \Rightarrow MS/VI \tag{8.21}$$

$$R_1 = F_1(H) + F_2(S) + F_3(S) \Rightarrow VI \tag{8.22}$$

where F_1 is the function of the data acquisition; F_2 is the function of the data processing; F_3 is the function of the user interface; S and H are the software and hardware realization respectively; TI is the traditional instrument; MS is the measuring system with GUI; VI is the virtual instrument.

However, it is not a complete set of basic models for measuring instruments. Let us consider some others. Let S be software in a broad sense, including programs for the general-purpose computer as well as software for embedded microprocessors and microcontrollers. Then three following models correspond to traditional instruments based on microprocessors or microcontrollers:

$$R_{1A} = F_1(H) + F_2(S) + F_3(H) \tag{8.23}$$

$$R_{1B} = F_1(S) + F_2(S) + F_3(H) \tag{8.24}$$

$$R_{1C} = F_1(S) + F_2(H) + F_3(H) \tag{8.25}$$

Equation (8.23) describes typical microprocessor instruments of the first generation, in which the microprocessor or the microcontroller executes only calculation functions for processing of the results. It is not the MS/VI, because in this case we have neither the software-based GUI, nor the virtual instrument.

The following two realizations (8.24) and (8.25) ask for attention. Equation (8.24) is true for the microcontroller or the microprocessor based measuring instruments of the new generation. In this case, the microcontroller or the microprocessor is used not only for calculation and control functions, but also for the realization of the measuring channel at the virtual level inside the functional-logical architecture. It means, that the A/D conversion is executed by the software. In this case, we have the so-called virtual measuring channel. It is built at the virtual level, inside the programmable computing power. In this case, the software can create a new metrological quality and new measuring functions. Apart from the frequency measurement, it is also possible to realize the measurement of the period, the pulse/pause duration, the period-to-pulse duration ratio and the duty-cycle without additional hardware and connections between the elements.

Microcontroller-based measuring instruments have the same features and advantages as the VI:

- software is the key

- low cost

- reusability

- flexible functionality.

In Equation (8.25) the data processing is performed by the hardware, that is necessary, for example, for the an increase in speed of arithmetic operations, connected with the processing of results, the Fourier transformation etc. In this case, the data processing is executed by functional expanders of microprocessor systems, the hardware ALU or co-processors.

In Equation (8.26), the data acquisition and the interface function (F_1 and F_3) are implemented by the software, the data processing (F_2) by the hardware:

$$R_{2A} = F_1(S) + F_2(H) + F_3(S) \tag{8.26}$$

This realization can be related to a virtual instrument or a measuring system with GUI, composed of the modern microcontroller-based measuring instruments.

Equation (8.27) also results in the same conclusion:

$$R_{3A} = F_1(S) + F_2(S) + F_3(S) \tag{8.27}$$

In these cases the measuring instrument can be qualified as a VI, when its hardware part and software part cannot operate separately as measuring instruments.

With allowance for that discussed above, the complete set of base models can be shown as:

$$
\left.
\begin{aligned}
R_1 &= F_1(H) + F_2(H) + F_3(H) \\
R_{1A} &= F_1(H) + F_2(S) + F_3(H) \\
R_{1B} &= F_1(S) + F_2(S) + F_3(H) \\
R_{1C} &= F_1(S) + F_2(H) + F_3(H)
\end{aligned}
\right\} \Rightarrow TI
$$

$$
\left.
\begin{aligned}
R_2 &= F_1(H) + F_2(H) + F_3(S) \\
R_{2A} &= F_1(S) + F_2(H) + F_3(S) \\
R_3 &= F_1(H) + F_2(S) + F_3(S) \\
R_{3A} &= F_1(S) + F_2(S) + F_3(S)
\end{aligned}
\right\} \Rightarrow MS/VI
\tag{8.28}
$$

where realizations R_i with alphanumeric indexes are the proposed models, adding to the existing set of base models. This complete set of base models of realizations for measuring instruments covers all instruments: the virtual instrument, traditional and microcontroller based intelligent measuring instruments, measuring systems and measuring instruments based on the hardwired logic.

Let us consider some examples of virtual instruments. First a virtual instrument for temperature measurements. It works with the smart temperature sensor SMT160-30 from SMARTEC [24]. This sensor has a duty-cycle output. The virtual thermometer is based on the virtual measuring channel. Therefore, the hardware is the minimum possible. All circuitry can be embedded into the COM or LPT-port connector. The block diagram of the sensor interfacing is shown in Figure 8.14.

Owing to the low-level software, the virtual channel is realized inside the microcontroller. One possible realizations is based on the 24-pin microcontroller Atmel AT89C2051, compatible with the MCS-51 microcontroller family (Intel). The duty-cycle-to-code conversion is carried out at a virtual level inside the microcontroller. In order to reach the high conversion speed, quasi-pipelining data processing is used in the microcontroller. In other words, the measurement, the calculation and the code conversion for transmission by the I/O interface are combined in time.

The graphical user interface is realized with the help of high-level software. The actual work was done using the software package ComponentWorks V2.0 [156], a

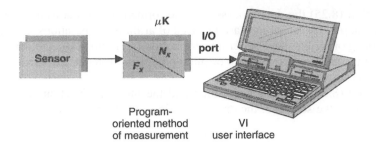

Figure 8.14   Block diagram of sensor interfacing

32-bit ActiveX (OLE) control for the test building and measurement, the analysis, and presentation systems in MS Visual Basic 6.0 environment. The screen display of one possible graphical user interface is shown in Figure 8.15. Controls allow the user to choose the accuracy (quantization error) or the time of measurement, informative parameters of the sensor and units. Each press of the 'Start' button on the front panel causes the beginning of the measuring cycle. The high-level PC software reads the data from the serial port and puts them in a file for consequent processing.

Technical performances are the following:

- temperature range, °C−45 . . . +130
- absolute accuracy, °C ±0.7

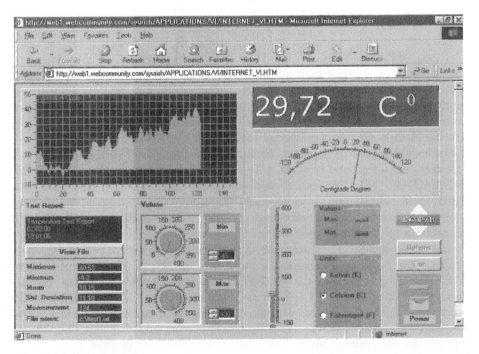

Figure 8.15   Screen display of graphical user interface for virtual instrument for smart temperature CMOS sensor

- relative error, % ±0.54

- linearity better than 0.2 °C

- direct connection to the LPT port

- cable length up to 20 m

- statistics calculation

- results file

- digital, analog and sound indications.

The minimum hardware requirements are: Pentium PC, 133 MHz, 16 MB RAM, and SVGA monitor. The operation system is Windows 95/98/Me.

A DAQ board for frequency/time parameters was also proposed based on the above-mentioned method of measurement. The block diagram of the proposed low-cost DAQ board is shown in Figure 8.16. The DAQ board uses the 82C54 (Intel) counter/timers for time-related functions. Although full integration of all elements as a single-chip implementation is an elegant approach, in most cases it may not be viable for low-cost DAQ boards.

Another example is a virtual instrument for pressure sensors. Despite the broad spread of pressure sensors with frequency output many sensors with voltage output continue to be used for different industrial applications. The ADC conversion, the long-distance transmission in a high-noise industrial environment are bottlenecks for the conventional solution of interfacing sensors with voltage readout with PCs. Also, the use of a traditional data-capturing method using an analog-to-digital card is more expensive. For transmission of the analog signal from the sensor to the PC, commensurable with distances as those ensured by the serial interface RS-232, it is necessary to

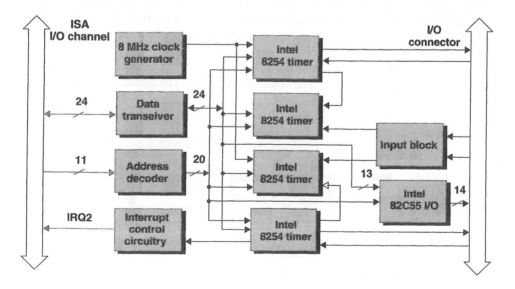

Figure 8.16 DAQ board block diagram

use additional measures. It means, at first, the use of additional devices for the signal condition, amplification and filtering. All these problems are specific for long distance analog signal transmission from the sensor to the PC. However, these problems are absent in the low-cost effective interfacing of sensors with voltage readout to PCs by digital signal transmission with the help of RS-232 interface.

The implementation of processing electronics in one chip together with the sensor element is an elegant and rather preferable engineering solution and is achieved by modern smart silicon pressure sensors. However, the combination of hybrid-integrated processing electronics with advanced processing methods in many cases allows magnificent technical and metrological performances for a shorter time-to-market period without additional expenditures for expensive CAD tools and the long smart sensor design process. For implementation of pressure smart sensors with hybrid-integrated processing electronics, hardware minimization is necessary to reach a reasonable price and high reliability. Besides, conversion methods used by the processing electronics must assure the high metrology performances, first high accuracy up to 0.01−0.005% proper to modern smart pressure sensors.

One possible solutions is based on the data-capturing method using a voltage-to-frequency-to-code conversion. The block diagram of the smart pressure sensor is shown in Figure 8.17.

The voltage-to-frequency converter provides unique characteristics when used as an analog-to-digital converter. Its excellent accuracy, linearity and integrating input characteristics often provide performance attributes unattainable with other converter types. In order to achieve the optimal speed/accuracy trade-offs, the program-oriented method of the dependent count is used for the frequency-to-code conversion as well as the quasi-pipelining data processing. The latter means that the measurement, calculation and code conversion for transmission by the RS-232/485 interface are combined in the same time. The frequency-to-code conversion is realized inside the functional-logical architecture of the microcontroller at a virtual level.

The technical realization of the hybrid-integrated processing electronics has been made for a pressure sensor with linear characteristics and 0−10 V full-scale voltage range. The circuit diagram of the voltage-to-frequency-to-code converter is shown in Figure 8.18.

Because this circuit diagram is universal for many analog sensors with the above-mentioned output voltage range, let us describe it in detail. The circuit works as follows.

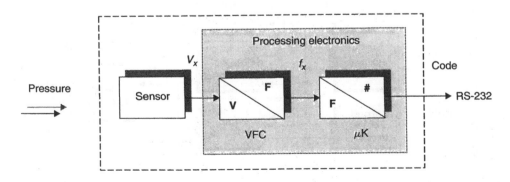

Figure 8.17   Smart pressure sensor

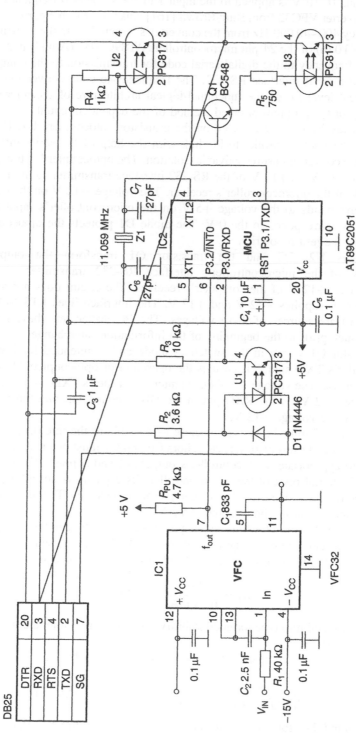

Figure 8.18 Voltage-to-frequency-to-code converter circuit diagram

An input signal 0–10 V is applied to the input 1 of the voltage-to-frequency converter IC1. The converter VFC32 from Burr-Brown [161] was used for this purpose. Impulses with frequency 400–40 000 Hz from the converter's output 7 go to the interrupt request input P3.2/INT0 (6) of the 24-pin microcontrollers AT89C2051 (Atmel) IC2. The latter converts the frequency to the digital serial code for transmission to the computer with the help of the serial interface RS-232. The frequency-to-code conversion is carried out at a virtual level inside the functional-logical architecture of the microcontroller with the help of the program-oriented method of the dependent count.

The signal level converter is based on the transistor optocouplers U1, U2 and U3, intended for the microcontroller-to-PC link with the help of the serial interface RS-232. The same converter ensures galvanic isolation. The optocoupler U1 transforms the output signal −12 V . . . +12 V of the RS-232 interface transmitter to the input signal +5 V . . . 0 V of the microcontroller's receiver. The voltage −12 V on the connector's pin TXD corresponds to the voltage +5 V on the microcontroller's input RXD, the voltage +12 V corresponds to the 0 V. The diode D1 protects the optocoupler from the voltage of the return polarity.

Optocouplers U2, U3 and the transistor Q1 transform the output signal 0 V . . . +5 V of the microcontroller's interface RS-232 transmitter to the input signal −12 V . . . +12 V of the computer's receiver. The circuit does not require any additional power supplies −12 V and +12 V. The interface lines RTS (+12 V) and DTR (−12 V) are sources of these voltages. The PC driver sets these lines to the appropriate state prior to the beginning of the information interchange.

The transistor Q1 works in the switching mode and is intended for inverting the microcontroller's TXD signal. It ensures the operation of optocouplers U2 and U3 in the opposite phase. The voltage +5 V on the microcontroller's TXD output corresponds to the voltage −12 V on the RXD pin in the DB25 connector, and the voltage 0 V corresponds to the voltage +12 V.

The capacitor C5 must be located as close as possible to the microcontroller's pins 10 and 20. The quartz resonator Z1, capacitors C6 and C7 must be also placed as close as possible to appropriate microcontroller's outputs (4 and 5 pins). The capacitor C4 is used for the initial reset of the microcontroller. Its capacity is not a bottleneck and can be varied over a wide range (from some μF up to tens μF). The operation voltage of the capacitor should not be less than +6.3 V.

The capacitor C3 is used for filtering the power supply voltage in the output cascade. A non-polar capacitor with voltage more than +25 V should be used as C3.

For $V_{FS} = 10$ V full-scale input, $R_1 = 40$ kΩ input resistor is chosen:

$$R_1 = \frac{V_{FS}}{0.25 \text{ mA}} = 40 \text{ k}\Omega$$

R_1 should be a metal film type for good stability. Manufacturing tolerances can produce approximately ±10%.

The full-scale output frequency (0–40 kHz) is determined by C1:

$$C_1 = \frac{V_{IN}}{7.5 \cdot f_{out} \cdot R_1} = \frac{10}{7.5 \cdot 40\,000 \cdot 40\,000} = 833.33 \text{ pF}$$

Any variation in C1 — tolerance, temperature drift, ageing — directly affects the output frequency. Ceramic NPO or silver-mica types are a good choice.

The value of the integrating capacitor, C2, does not directly influence the output frequency, but its value must be chosen within certain bounds. According to [161] this value is C2 = 2.5 nF. If this value is made too low, the integrator output voltage can exceed its linear output swing, resulting in a non-linear response. Use of C2 larger values is acceptable. Polycarbonate and other film capacitors are generally excellent. Many ceramic types are adequate, but some low-voltage ceramic capacitor types may degrade the non-linearity. Electrolytic types are not recommended.

Screen displays of the graphical user interface for pressure sensors are shown in Figure 8.19.

The high-level PC software reads the data from the serial port and puts them in a specific file for the consequent processing of results. It allows choice of the appropriate

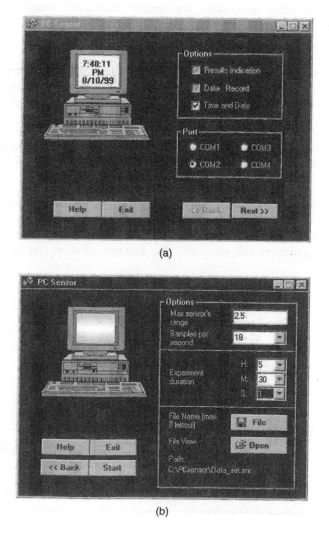

(a)

(b)

Figure 8.19   Screen displays of graphical user interface for pressure sensors

COM-port, the maximum conversion range, the number of samples per second, the experiment, etc. Results are saved in a file on the hard disk drive in ASCII text format.

The experimental results are shown in Figure 8.20 The aim of the experimental research was the determination of the total error for the frequency-to-code conversion in a specified measuring range of frequencies from 400 up to 40 000 Hz. The obtained results are the following: (1) With probability 97% the total error does not exceed the value 0.0064% in all specified measuring ranges of frequencies and is less than the specified relative quantization error δ; (2) The total error can be neglected, as it is two orders less than the nonlinearity error of the voltage-to-frequency converter and three to four orders less than the sensor's error.

The dependence of the total error of measurement on frequencies is explained by the availability of the calculation error of the results. It depends on features of floating point arithmetics used in the microcontroller, namely, from operands.

There are some other advantages for sensor interfacing using the voltage-to-frequency-to-code conversion:

- Voltage-to-frequency converters provide a simple, low-cost alternative to A/D converters. The frequency output is easily isolated, transmitted or recorded. It can be interfaced to many commonly used microcontrollers and microprocessors.

- The voltage-to-frequency converter's integrating input properties, the excellent accuracy and the low nonlinearity provide performance attributes unattainable with other converter types. All of this makes them ideal for high-noise industrial environments.

- Since an analog quantity represented as a frequency is inherently a serial data stream, it is easily handled in large multichannel systems. The frequency information can be transmitted over long lines with excellent noise immunity using low-cost digital line transmitters and receivers. The voltage isolation can be accomplished

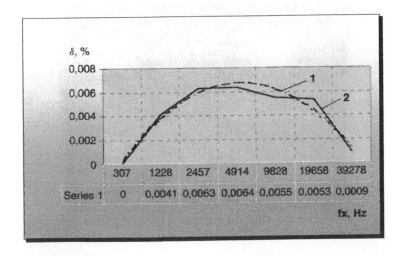

Series 1	0	0,0041	0,0063	0,0064	0,0055	0,0053	0,0009

Figure 8.20 Total error of conversion $\delta = \phi(f_x)$ (1 = trend line; 2 = experimental results)

with low-cost optical couplers or transformers without loss in accuracy. Many channels of frequency data can be efficiently steered using simple hardware, avoiding expensive analog multiplexing circuitry.

- Like a dual-slope A/D converter, the VFC possesses a true integrating input. While a successive approximation the A/D converter takes a 'snapshot' in time, making it susceptible to noise peaks, the VFC's input is constantly integrating, smoothing the effects of noise or varying input signals.

- The minimum hardware and the size due to the program-oriented method of measurement (conversion) make it possible to place the circuitry inside sensors or even inside the RS-232 cable connector.

- The proposed solution has universality. First, it has a user-selected voltage input range (±Supply). Second, the high accuracy of the frequency-to-code conversion (up to 0.001%). The error of such a conversion can be neglected in the measuring channel. This is not true for the analog-to-digital conversion. The ADC's error is commensurable with the sensor's error, especially if we are using modern high precision sensors with relative error up to 0.01%.

The last but not least example of the virtual instrument in this book is the virtual tachometric system. Modern turbogenerators in power stations are arranged with automatic control systems of the rotation speed. They ensure the opportunity of disconnecting appropriate valves of the turbine and suspend the arrival of a pair when emergencies arise. The control system is also supplemented by circuit-breakers in case the control system cuts outs of the operation. The reliable and exact operation of such systems eliminates the unplanned stopping of turbines and certain emergencies, which can destroy the turbine. Therefore, stringent requirements for the reliability and accuracy of measurements are imposed to the tuning and the test of such control systems. It is connected, from the one hand, with inadmissibility of emergencies and from the other hand with increasing the efficiency of a turbogenerator.

This task can be solved by designing a specialized virtual tachometric system. The measuring instrument is based on the program-oriented ratiometric counting method of measurement for the rotational speed. This method also satisfies modern technical requirements by using considerably less hardware and simple realization. It is achieved by the use of the measuring channel inside the functional-logical architecture of the one-chip microcontroller at the virtual level. Apart from measuring instruments, the tachometric system includes the semiconductor active sensor of the rotation speed [52], two sensors of the circuit-breaker's firing-pins, the interfacing block for a rotation speed sensor with the turbogenerator's shaft and the set of connecting cables. The architecture of such a system depends on the type of turbine.

Like a conventional tachometer (Figure 8.21) the measuring information is displayed on the virtual digital and analog indicators of the virtual tachometer. The 25 fixed values of the rotation speed by the speed-up of the turbine, as well as the moment of operation and loosening of the circuit-breaker's firing-pins, are fixed on the virtual LED panel. The LED indication is accompanied by sound signals with various tones and time duration.

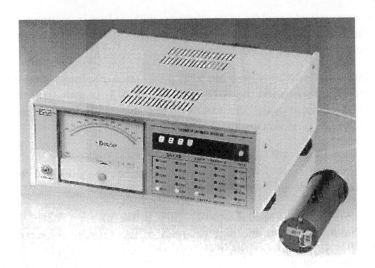

Figure 8.21   Conventional tachometer and firing-pins sensor

The sensor has a wide frequency range $(0\ldots 50\,000$ rpm), small overall dimensions, mass and power consumption. It best approaches the determination of the shaft state 'Shaft is stopped' $(n_x \leq 12$ rpm).

The measurand rotation speed is given by

$$n_x = \left(\frac{N_1}{N_2}\right) \cdot f_0 \cdot \frac{60}{Z} \qquad (8.29)$$

Since for the particular tachometric system the factor $f_0 \cdot 60/Z$ is constant, finally we have

$$n_x = \frac{N_1}{N_2} \cdot k_n \qquad (8.30)$$

The sensors of the firing-pins operation of the circuit-breaker are based on the principle of the contacts in a magnetic field. They have a compact construction, which takes into account features of the turbogenerator and reliably fasten onto its stator. These sensors automatize the process of fixing of its firing-pins position.

The flow diagram of the tuning is shown in Figure 8.22. The measurement instrument functions as follows. After power up, the initialization of the system for realization of analog, digital and sound indications takes place. After that, automatic self-testing of the instrument occurs. At the test ending, the tachometer will be automatically turned into the mode of measurement of frequency, which comes from the rotation speed sensor. After that, the program-oriented method of measurement for frequency is executed. At the end of the first cycle of measurement, the subroutines of multibyte multiplying are executed with aim of result scaling. After that, the subroutine multibyte division for calculation of the result of the indirect measurement of the rotation speed according to Equation (8.29) and its indication in 'rpm' units are executed. The binary code of the measurement result is transformed into the binary-coded decimal code unpacked format. Then a check is carried out on whether the current cycle of the measurement is the first. If so, the result of measurement is indicated on the LED indicator, if not,

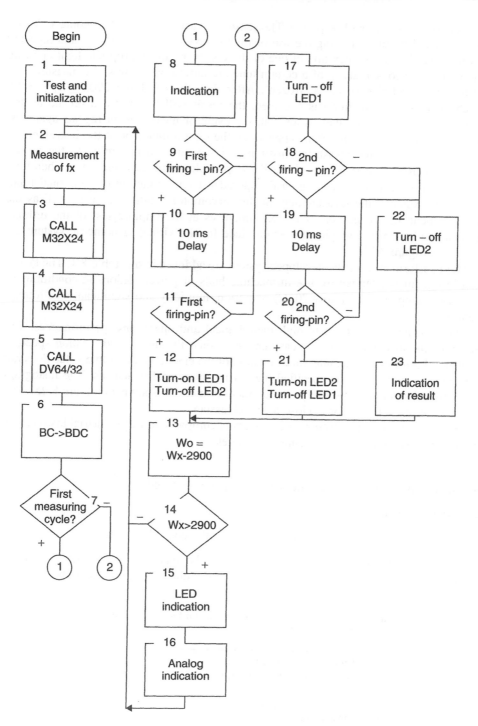

Figure 8.22 Flow chart of tuning algorithm

sensors of firing-pins will be polled. The sensor contact chatter by their contact closure is eliminated by programming the use of time delays of duration 10 ms and repeated sensor polling. By the operation of one of these sensors, the appropriate light-emitting diode and an acoustic alarm of a certain tone are turned on. At this, the measurement result of the rotation speed is 'frozen'. After that, the condition $2900 \leq n_x \leq 3400$ is checked. If this condition is executed, the result will be indicated on the analog indicator and the LED panel. After that, the cycle of measurement repeats again. At $n_x \leq 12$ rpm, the message '*Stop*' appears on the digital indicator of the instrument.

The software of the measuring instrument is constructed according to the hierarchical principle. The structure of the developed algorithm has openness. Therefore, the measuring instrument can be easily upgraded by introduction of additional functions, for example, the measurement of the servomotor handling time as well as the measurement and generation of the test impulses at the impulse test of the turbine. Screen displays of the graphical user interface for the virtual tachometric system are shown in Figure 8.23.

Main performances of the developed system are adduced in the Table 8.1. The tachometric system is intended for use in machine halls of power stations in conditions of increased temperature (up to $+70\,°C$), vibration, humidity and significant electromagnetic fields.

Experiments have confirmed high metrological and operational performances. The graph of the absolute error of measurement against the measurand rotation speed is shown in Figure 8.24. The interval evaluation of the error on an absolute value does not exceed 0.57 rpm in all the specified measuring range of frequencies. The histogram of distribution of the absolute error of measurement is shown in Figure 8.25. The distribution law of an error is close to the Gaussian.

This specialized virtual tachometric system in comparison with industrial instruments of similar assignment has following advantages:

- increased accuracy of measurement and operating speed

- extended functionality

- full automatization of the tuning process for the control system

- the possibility of easy substitution of sensors, the change of gradation of the rotor-modulator and the possibility of use with any models of turbogenerators

Table 8.1   Main performances of virtual tachometric system

Parameter	Range
Specified measuring range, rpm	10–9999
Absolute error of measurement, rpm	0.57
Relative error, %	0.05
Accuracy class of analog indication, %	0.5
Time of measurement, sec	0.25
Number of fixed points for the LED indication	25
Input signal, V	0.035–24
Current consumption, mA	600

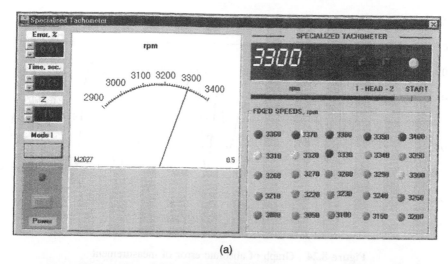

(a)

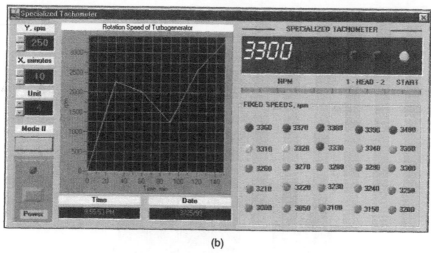

(b)

Figure 8.23   Virtual specialized tachometer in different mode of measurement

- minimum possible hardware and high adaptability for manufacturing of such an instrument.

The use of this tachometric system gives a real effect of electric power saving due to more precise and fast tuning of the control system of the rotation speed of a turbogenerator's rotor as well as during the production and the turbine service.

8.7   Estimation of Uncertainty for Virtual Instruments

The usual practice in the design and creation of modern intelligent measuring systems based on virtual instruments is to purchase and use commercial sensors and trans- ducers, data acquisition boards, signal condition units, signal processing modules and

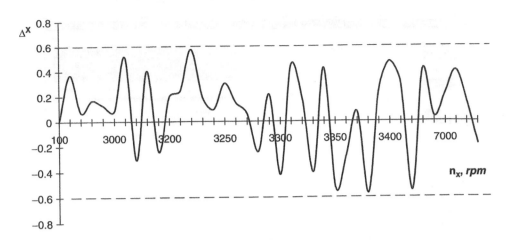

Figure 8.24   Graph of absolute error of measurement

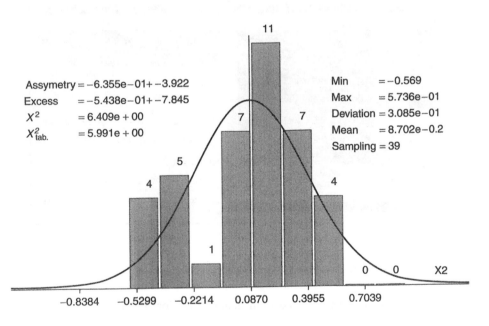

Figure 8.25   Histogram of distribution for absolute error of measurement

measurement software. In such a way, we can create a measuring system of any required configuration and functionality. Even if all components and blocks have certified metrological characteristics (specifications), there is always a problem of how to characterize the final metrological properties of the arranged system as a whole, especially if tailored for the user. Beside that in many practical applications, measurement procedures are rather complicated. In general, the uncertainty of the final measurement depends on many factors such as input values of the measurand and its error caused by non-ideal transformations, the data acquisition and processing algorithms, etc. and cannot be evaluated manually [159]. Nevertheless, the fundamental requirement of the general and legal metrology is that the accuracy of measurement results should be estimated and reported. It is especially vital for virtual instruments and sensor systems when parameters of modules and data acquisition algorithms are varied automatically or by the user and consequently the accuracy of results may be changed.

The general approach for the determination of the probability distribution of a summary error of measurement is reduced to the calculation of the summary distribution of a sum of independent random values and the application of methods from a probability theory. Such a method of summation is the most accurate, but rather labour consuming, as all probable likely characteristics and properties of summing-up errors must be taken into account. Such calculation requires additional initial information and must make some heuristic decisions during the calculation. Sometimes it causes serious difficulties. For example, the change from the standard deviation to the entropy or confidence values is the most difficult operation at the last stage of summation for errors. This operation assumes the determination of the frequency function form, as the entropy factor or the quantile multiplier of a resulting error is dependent on the probability distribution. However, classical techniques rely on the fact that the exact determination of the resulting distribution of the sum of all its components must be omitted.

The η-method of summation of an error's components, which considers the frequency function form on the basis of the excess η, is described in [162]. However, the excess does not determine unequivocally the type of probability distribution.

The uniform technique of the automated construction of the frequency function and the determination of the performance data for the resulting error of n components, which easily yielded to algorithmization and could be realized using computer software, do not exist. At the same time, the class of probability distributions for errors of measurements is well determined and all of them, except the Rayleigh distribution are met in practice by the solution of metrological tasks.

Let us consider the method of construction of the probability distribution and the determination of performances for a resulting error of n components with a given accuracy. This method is based on the use of Liapunov characteristic functions (LCF). In comparison with the method of the frequency function construction described in [163] in this method the piece-wise linear approximation of the probability distribution $W(\delta)$ with the automatic change of the discretization step dependent on the given accuracy is used instead of step approximation [164]. It obtains an optimum speed/accuracy ratio by the computer realization of the algorithm for this method. Such probability distributions as rectangular, triangular, trapezium and some others are approximated without error. The distribution, which is impossible to be precisely approximated by straight-line segments is approximated with the given accuracy. From piece-wise linear approximation, the function $W(\delta)$ can be presented as

$$W(\delta) = \sum_{i=1}^{n} (1(\delta_i) - 1(\delta_{i+1})) \cdot (A_i\delta + W_i\delta_{i+1} - W_{i+1}\delta_i), \qquad (8.31)$$

where A_i is the coefficient, which is calculated according to the following equation:

$$A_i = \frac{W_{i+1} - W_i}{\delta_{i+1} - \delta_i}; \qquad (8.32)$$

W_i, d_i are coordinates of the i-th point, where the distribution $W(\delta)$ is approximated; $1[.]$ is the unit Heavyside function; n is the number of points, which approximates the distribution $W(\delta)$ of the random error δ.

Having calculated the LCF $\theta(j\omega)$ of the distribution $W(\delta)$, we have

$$\theta(j\omega) = M(e^{j\omega\delta}) = \int_{-\infty}^{+\infty} e^{j\omega\delta} W(\delta)d\delta = \theta_1(j\omega) + j\theta_2(j\omega), \qquad (8.33)$$

where $\theta_1(j\omega)$, $\theta_2(j\omega)$ are real and imaginary parts of the required LCF accordingly. We shall determine θ_1 and θ_2 separately:

$$\theta_1 = \int_{-\infty}^{+\infty} W(\delta) \cos \omega\delta d\delta = \sum_{i=1}^{n} A_i(\omega^{-2} \cos \omega\delta_{i+1} + \delta_{i+1}\omega^{-1} \sin \omega\delta_{i+1} - \omega^{-2} \cos \omega\delta_i$$

$$- \omega^{-1}\delta_i \sin \omega\delta_i) + (W_i\delta_{i+1} - W_{i+1}\delta_i)\omega^{-1}(\sin \omega\delta_{i+1} - \sin \omega\delta_i). \qquad (8.34)$$

After elementary rearrangements of the composed sum and assuming that there are some points with $W = 0$ at the beginning and at the end of the frequency function curve, we have:

$$\theta_1(\omega) = -\frac{2}{\omega^2} \sum_{i=1}^{n} A_i \sin \frac{\omega(\delta_i + \delta_{i+1})}{2} + \frac{1}{\omega} \sum_{i=1}^{k} B_i \sin \omega\delta_i, \qquad (8.35)$$

where k is the number of first sort breaks in the frequency function curve $W(\delta)$; B_i is the coefficient, which can be calculated according to the formula

$$B_i = W_{i,1} - W_{i,2}, \qquad (8.36)$$

where $W_{i,1}$, $W_{i,2}$ are the left-hand and right-hand borders in the first sort break point δ_i accordingly:

$$W_{i,1} = \lim_{\delta \to \delta_i+0} W(\delta) \qquad (8.37)$$

$$W_{i,2} = \lim_{\delta \to \delta_i-0} W(\delta) \qquad (8.38)$$

Let us determine the imaginary part of the LCF:

$$\theta_2 = \int_{-\infty}^{+\infty} W(\delta) \sin \omega\delta d\delta. \qquad (8.39)$$

After similar transformations:

$$\theta_2(\omega) = \frac{1}{\omega^2} \sum_{i=1}^{n} A_i \sin \frac{\omega(\delta_i - \delta_{i+1})}{2} \cdot \cos \frac{\omega(\delta_i - \delta_{i+1})}{2} + \frac{1}{\omega} \sum_{i=1}^{k} B_i \cos \omega\delta_i. \qquad (8.40)$$

The real part $\theta_1(j\omega)$ of the characteristic function $\theta(j\omega)$ is even, and the imaginary part $\theta_2(j\omega)$ is the odd function of the frequency ω. Therefore, it is enough for the $\theta(j\omega)$ calculation in the positive area ($\omega > 0$), and in the negative area, it is possible to calculate the LCF according to the following equation:

$$\theta(j(-\omega)) = \theta_1(\omega) - j\theta_2(\omega) \qquad (8.41)$$

For the computer representation of the LCF it is also necessary to approximate the latter by the piece-wise line and to store initial values of real and imaginary parts. The approximation is executed automatically with the beforehand given accuracy ε. Thus, the spectrum is limited by the frequency ω_{max}, at which

$$\begin{cases} |\theta(j\omega_{max})| \leq \theta_{max}\delta_0 \\ |\theta'(j\omega_{max})| \leq \theta'_{max}\delta_0 \end{cases} \tag{8.42}$$

where ω_{max} is the maximum value of frequency ω from the range $[0, \omega_{max}]$, within the limits of which the value of LCF is calculated; δ_0 is the calculation (approximation) accuracy of the $\theta(j\omega)$; $|\theta_{max}|$ is the maximum value of the LCF module; $|\theta'_{max}|$ is the maximum value of the derivative module of the LCF.

If LCFs are known (the LCF of error's components are in the PC's memory point-by-point), it is necessary to multiply them together and to approximate the resulting LCF by the piece-wise line with accuracy δ_0. Hence, it is possible to present the summary LCF $\theta_\Sigma(j\omega)$ as

$$\theta_\Sigma(j\omega) = \sum_{i=1}^{n}(1(\omega_i) - 1(\omega_{i+1})) \cdot \left(\frac{\theta_{i+1} - \theta_i}{\omega_{i+1} - \omega_i} \cdot \omega + \theta_i\omega_{i+1} - \theta_{i+1}\omega_i\right). \tag{8.43}$$

Having executed the return transformation with allowance that the real part of any LCF is the even function of ω, and the imaginary part is the odd function of ω, we obtain the probability distribution of the resulting random error $W_\Sigma(\delta)$:

$$W_\Sigma(\delta) = \frac{1}{2\pi}\int_{-\infty}^{+\infty}\theta_1(\omega)\cos\omega\delta d\omega + \frac{1}{2\pi}\int_{-\infty}^{+\infty}\theta_2(\omega)\cos\omega\delta d\omega$$

$$+ \frac{j}{2\pi}\int_{-\infty}^{+\infty}\theta_1(\omega)\sin\omega\delta d\omega + \frac{1}{2\pi}\int_{-\infty}^{+\infty}\theta_2(\omega)\sin\omega\delta d\omega \tag{8.44}$$

Taking into account, that

$$\int_{-\infty}^{+\infty} A(\omega)d\omega = 0, \tag{8.45}$$

where $A(\omega)$ is the odd function of ω

$$\int_{-\infty}^{+\infty} B(\omega)d\omega = 2\int_{0}^{+\infty} B(\omega)d\omega \tag{8.46}$$

and $B(\omega)$ is the even function of ω, we have

$$W_\Sigma(\delta) = \frac{1}{2\pi}\int_{-\infty}^{+\infty}\theta_1(\omega)\cos\delta\omega d\omega - \frac{1}{2\pi}\int_{-\infty}^{+\infty}\theta_2(\omega)\sin\delta\omega d\omega. \tag{8.47}$$

The equation has the similar structure to that considered above. Therefore, the final result of transformation:

$$W_\Sigma(\delta) = \frac{1}{\pi\delta}\sum_{i=1}^{n} C_i \cdot \cos\omega_i\delta D_i \sin\omega_i\delta, \tag{8.48}$$

where n is the number of points of the $W_\Sigma(\delta)$ approximation; C_i and D_i are coefficients, which are calculated as following:

$$C_i = \frac{P_i - P_{i-1}}{\omega_i - \omega_{i-1}} - \frac{P_{i+1} - P_i}{\omega_{i+1} - \omega_i} \tag{8.49}$$

$$D_i = \frac{R_i - R_{i-1}}{\omega_i - \omega_{i-1}} - \frac{R_{i+1} - R_i}{\omega_{i+1} - \omega_i} \tag{8.50}$$

where P_i, R_i are real and imaginary parts of the summary LCF of $\theta_\Sigma(j\omega)$ in the point ω_i. For the determination of numerical characteristics of the distribution on its LCF, we shall use basic statistical calculations, in which we shall replace derivatives by final differences. Then the logarithmic LCF and its four derivatives will have the following form:

$$\theta_{l,i} = \ln(\theta_i), \quad i = \overline{1, \dots, 5} \tag{8.51}$$

$$\theta'_{l,i} = \frac{\theta_{l,i+1} - \theta_{l,i}}{\omega_{i+1} - \omega_i}, \quad i = \overline{1, \dots, 4} \tag{8.52}$$

$$\theta''_{l,i} = \frac{\theta'_{l,i+1} - \theta'_{l,i}}{\omega_{i+1} - \omega_i}, \quad i = \overline{1, \dots, 3} \tag{8.53}$$

$$\theta'''_{l,i} = \frac{\theta''_{l,i+1} - \theta''_{l,i}}{\omega_{i+1} - \omega_i}, \quad i = \overline{1, \dots, 2} \tag{8.54}$$

$$\theta^{IV}_{l,i} = \frac{\theta'''_{l,i+1} - \theta'''_{l,i}}{\omega_{i+1} - \omega_i}. \tag{8.55}$$

Equations for the numerical characteristics determination of the error with allowance for Equations (8.51–8.55) have the following form:

$$M(\delta) = i^{-1}\theta'_{l,i}, \tag{8.56}$$

$$D(\delta) = -\theta''_{l,i}, \tag{8.57}$$

$$\sigma = \sqrt{D}, \tag{8.58}$$

$$K_{\text{asymmetry}} = \theta'''_{l,i}[\theta''_{l,i}]^{-3/2}, \tag{8.59}$$

$$K_{\text{excess}} = \theta^{IV}_{l,i}[\theta''_{l,i}]^{-2} \tag{8.60}$$

For realization of the automatic change for the discretization step by the calculation of any function (the frequency function or the LCF), its value is calculated in three points. Arguments of these points will be distinguished from one another by the discretization step h. The absolute accuracy of calculations is determined as the module of the difference between the value of function in the central point and in the middle of the straight-line segment, which connects its two extreme points of approximation. Then the absolute accuracy of calculations is determined by the formula

$$\Delta E = \left| \frac{(2F_2 - F_1 - F_3)}{h} \right|, \tag{8.61}$$

where F_1, F_2, F_3 are values of the function in the first second and third points of approximation accordingly; h is the discretization step. The discretization step can be decreased or increased depending on the required calculation accuracy.

This method can be formalized rather easily. Based on the offered algorithm a software package for the automation construction of the probability distribution and determination of performances for the resulting error of n components was developed. The friendly interface of the software package enters the accuracy ε in %, chooses the distribution law for the components and inputs their characteristics in the dialogue mode. The real calculation accuracy will be $\sqrt{n}$ times more than the given, where n is the number of components of the resulting error. Apart from the Gaussian, triangular, trapezium, uniform, anti-modal 1 and anti-modal 2 frequency functions, the software package also uses the Erlang and the linear-broken distribution. The latter holds a special place, as it represents any probability distribution, which can be represented by the final sequence of points, connected by straight-line segments. In this case, all points of bends are entered consistently, from δ_{min} up to δ_{max}.

The software package supplies the following output information:

- numerical characteristics of errors for the entered component as well as the resulting error (the mean, the dispersion, the standard deviation, coefficients of asymmetry and excess, the confidence interval at any probability, up to $P_{max} = 0.9999$)

- the frequency function curve

- amplitude-frequency and phase-frequency characteristics for the LCF.

Numerical characteristics and the composition curve of the probability distribution give more complete and authentic characteristics about metrological performances of measuring instruments and systems.

We shall consider an example of the calculation of an error for the measuring channel and compare the results, from a hand-operated technique and the novel automated method of summation for errors. Calculations were performed on the initial data from the well-known (classical) example.

It is required to calculate the resulting error for a measuring channel, consisting of serial units (a sensor, an amplifier and a recorder). An error of zero of the measuring channel with the analog recorder includes four components: the recorder's error σ_{rec}, the intrinsic sensor's error σ_{sen}, the temperature sensor's error σ_t and the summary error of two rigid correlation components: the amplifier zero displacement error by the temperature fluctuation and the temperature error of the recorder $\sigma_{t(ampl+rec)}$. It is possible to neglect the two last components (for consistency of comparison of two techniques), because one is 9, and the other is 13.5 times less than the recorder's error. Probability distributions of the components are uniform (Figure 8.26). Comparative results of calculation are adduced in Table 8.2.

Numerical characteristics of the resulting distribution law from the software package are shown in Figure 8.27. The resulting distribution of two components σ_{reg} and σ_{sen} is shown in Figure 8.28. If the two omitted components were not neglected, the resulting error would be equal to 0.26%. The channel error with the digital recorder includes the same error of the sensor and the error of the digital voltmeter. It is also possible to neglect the temperature sensor's error and the amplifier's displacement zero error

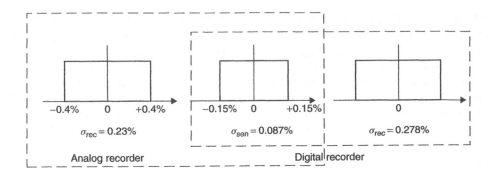

Figure 8.26   Probability distributions of components

Table 8.2   Comparative results of calculation for channel with analog recorder

Error, %	Distribution law for the component	Resulting error, % (at P = 0.98)	
		Standard method	New method (E = 0.01%)
$\sigma_{rec} = 0.23$	uniform		
		0.25	0.28
$\sigma_{sen} = 0.087$	uniform		
Summary frequency function		Not defined	Trapezium

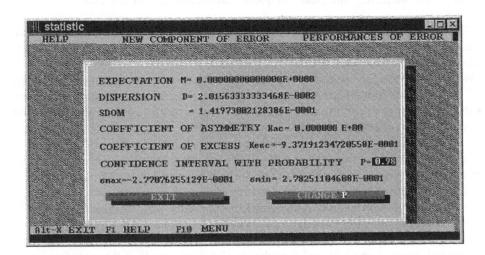

Figure 8.27   Numerical characteristics of resulting distribution law

because of the temperature fluctuation. It is fair according to the rule to neglect small components by the summation of errors. Comparative results of calculation are shown in Table 8.3.

The novel method of summation of errors allows an increase in the accuracy of determination of numerical characteristics for the resulting error up to 10–11%, completely

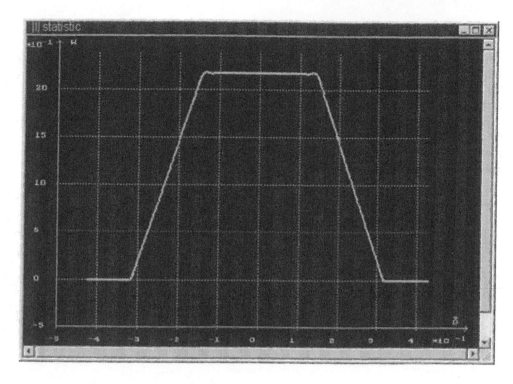

Figure 8.28   Resulting distribution of two components σ_{reg} and σ_{sen}

Table 8.3   Comparative results of calculation for channel with digital recorder

Error, %	Distribution law of the component	Resulting error, % (at P = 0.98)	
		Standard method	New method (E = 0.01%)
$\sigma_{rec} = 0.278$	uniform		
		0.291	0.327
$\sigma_{sen} = 0.087$	uniform		
Summary frequency function		Not defined	Trapezium*

*Numerical characteristics of the resulting probability distribution, received from the software package are: $M = 0$; $D = 2.83 \times 10^{-2}$; SDOM $= 0.168$; $K_{asymmetry} = 0$; $K_{excess} = -1$; confidence interval $-3.211 \times 10^{-1} \leq X \leq 3.267 \times 10^{-1}$ at P = 0.98%.

automatizes the calculation, the visualization and the registration of the distribution law, and consequently, an increase in the labour productivity by the individual certification of each measuring instrument.

The results can be fixed on the basis of computer techniques for the analysis and synthesis of probability distributions by the design of measuring instruments and the determination of the accuracy class by the certification of measuring instruments and systems. The offered technique can also be useful for problems of the automatic on-line evaluation of uncertainties of the final measurement result in virtual measuring instruments as well as for other similar problems, connected with the analysis and research of results of various measurements.

Summary

The method of the dependent count for the frequency-to-code converter easily realizes smart sensor systems with the time- and space-division frequency channelling. Due to the non-redundant conversion time the method can be used in different time-critical applications, for example, the ABS as well as in multiparameters sensors.

Due to the 'virtual instrumentation revolution', we can create a measuring system of any required configuration and functionality. In order to estimate final metrological properties of the arranged system it is necessary to use an advanced technique for the error estimation.

9

SMART SENSOR DESIGN AT SOFTWARE LEVEL

Because modern smart sensors include as a rule a microcontroller, the problem of software level design has arisen. A microcontroller core software level design allows minimization of the chip area, in turn, the software low-power design technique allows reduction of the power consumption. Both are very important for smart sensors and can improve sensor characteristics without additional labour expenses and time. However, until now such a design was realized as a rule, heuristically. Only some of its aspects are reflected in the literature and included in CAD tools. We hope that in this connection, the material described below, the set of 'hints' and recommendations will be useful for readers.

9.1 Microcontroller Core for Smart Sensors

The measuring instrumentation was, and remains, one of the main applications of embedded microprocessors and microcontrollers. The well-known efficient family of microcontrollers MCS-51 (Intel), which appeared at the beginning of the 1980s, continues to be used actively for different smart sensor applications. This is explained by the following:

1. the low cost

2. the availability of inexpensive debugging tools

3. the availability of well-trained engineers.

A microcontroller generally comprises two main parts–a central processing unit (CPU) often called a core, and a peripheral part–different peripheral modules connected to the CPU. The core can also be used as a control unit with other peripherals in an ASIC or a system on a chip. As a standard library cell the 8051 compatible microcontrollers exist in many CAD tools for the VLSI and the ASIC. Over a 20-year period dozens of modifications to the classical microcontroller 8051 with the Harvard architecture have been made. Due to modern microelectronics, the maximum clock frequency of such microcontrollers has increased to 50 MHz. The microcontrollers of the MCS-51 family are general-purpose microcontrollers, and their instruction set has the essential redundancy for metering applications. The instruction set of a standard peripheral unit is often not used completely in a design. The underlying idea of the smart sensor design approach is to shorten the instruction set, to reduce the processor area and improve functional performances by including other system parts in the same chip.

Static analysis was used to reduce the silicon area [165]. It revealed possibility of eliminating some instructions because they were not used at all in the sensor (metering) applications [166]. In general, the software includes the test software, the PCM and arithmetic procedures. Let us consider, for example, the rotational speed sensor with an embedded microcontroller core. The application is a tachometric system. The total number of program instructions is 722. Table 9.1 shows the percentage of used instructions in the measuring instrument program after static analysis was performed. Instructions are organized as groups.

The MOV group, for instance, includes all addressing modes for the MOVE instruction available in the 8051 microcontroller. As can be seen, nine instructions are responsible for more than 80% of the code. There are 27 total groups of instructions.

Table 9.1 Groups of often used instructions for metering applications

No.	Type	Number of instructions	%
1.	MOV	249	34.49
2.	LCALL (ACALL)	92	12.74
3.	MOVX	58	8.03
4.	NOP	39	5.4
5.	INC	37	5.13
6.	AJMP (LJMP)	31	4.29
7.	DJNZ	30	4.16
8.	ANL	24	6.65
9.	RET	24	6.65
			80.89%
10.	ORL	20	2.77
11.	CLR	19	2.63
12.	XCH	18	2.49
13.	JB	16	2.22
14.	JZ	11	1.52
15.	DEC	9	1.25
16.	CPL	7	0.97
17.	ADD	7	0.97
18.	JNZ	6	0.83
19.	ADDC	6	0.83
20.	RL	4	0.55
21.	RR	4	0.55
22.	JNC	4	0.55
23.	DA	2	0.28
24.	RETI	2	0.28
25.	RLC	1	0.14
26.	RRC	1	0.14
27.	XRL	1	0.14
TOTAL: 27 groups			100%

The information that is really needed in order to reduce the chip area is the number of never-used instructions obtained during the static analysis. If an instruction set was not used by the processor, there would be no need to include it in the processor hardware. These instructions can be eliminated from the processor control area, saving space for other important hardware. In this case, we expect an area saving of almost a third in comparison with the original processor core. The saved chip area can be used for additional 8253 timer/counters with the addressing logic. Static analysis revealed that seven types of instructions from a possible nine types, are the same as for the induction motor control software [165].

9.2   Low-Power Design Technique for Embedded Microcontrollers

The cost of microcontrollers and other electronics building blocks such as voltage-to-frequency and frequency-to-code converters is continuously decreasing and this allows many new possibilities for smart sensors. Due to embedded microcontrollers a great deal of local signal conditioning can be taken over by the processor.

A modern microcontroller-enhanced smart sensor could be composed of a sensing element, a signal-processing part, generating, for example, the frequency–time-domain signal and a microcontroller. All these parts might be joined on one single chip forming a smart sensor. In such a promising case, the need for low-power consumption smart sensors with embedded microcontrollers is becoming increasingly urgent, especially for different wireless and human-body implemented sensors.

The critical importance of low-power circuits has attracted the efforts of several researchers and designers in the last few years. However, despite the wealth of results the problem of minimizing the power consumption for the embedded microcontroller system is open and partially unsolved, mainly because of the lack of a consistent and complete effort in addressing the problem from a global point of view, starting from the instruction (software) level to optimizing compilers, which could identify hot spots and better direct the optimization effort.

Even though microcontroller tasks for smart sensors are usually fixed for particular applications, the software and algorithm optimization provide considerable potential for an additional reduction of power consumption, particularly of program-oriented methods of conversion are used. The algorithmic optimization has resulted in using the non-redundant reference frequency for the frequency-to-code conversion [114], described in Chapter 5.

An integrated framework for low-power design must provide the user with tools for the low-power synthesis at different levels of abstraction. In particular, the power consumption needs to be targeted at the specification, architectural, instruction, logic and physical levels. If the microprocessor or the microcontroller is not in the house-designed Application-Specific Instruction Processor (ASIP) or the redesigned core, the instruction set level becomes very important as it can allow additional power saving from 10 to 35% in the final design. It is the software that directs much of the activity of the hardware in a microcontroller and the digital signal processor based system. Consequently, the software can have a substantial impact on the power dissipation of a system [167]. From the point of view of the software low-power design, embedded

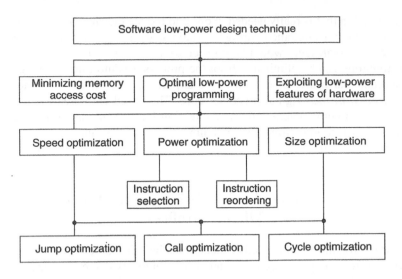

Figure 9.1   Software low-power design technique

microprocessors and microcontrollers must be programmed in a 'low-power program-ming style' [168].

The software low-power design technique encompasses the main problems: the optimal low-power programming, minimizing memory access cost and exploiting low-power features of the hardware [169] (Figure 9.1).

The compiler technology for the low power is based on the software optimization at the instruction level. It includes the following measures: speed, size and, directly, the power optimization. The latter is executed through special power-saving 'hints': instruction selection and reordering. These techniques need to be incorporated into auto-mated software development tools (optimizing compilers) to be used extensively. There are two levels of tools: tools that generate a code from an algorithm specification and the compiler level optimization. Although such tools exist for DSP applications [170], they do not yet exist for embedded microcontrollers. The compiler technology for low power appears to be further along than high-level tools partially because well-understood performance optimizations can adapt to energy minimization.

Considerable work has been done on compiler optimization to achieve the maximum performance, but little consideration has been given to optimization for low powers. Existing compilers allow the user to select between optimizing for code size and opti-mizing for performance; adding a third, user-selectable criterion to cause the compiler to optimize for power efficiency is quite feasible. Consequently, the development of optimum low-power programming methods and optimization at the instruction level is a high priority task.

The first main step towards software optimization for smart sensor low-power design is to estimate the power dissipation for a piece of a code. This has been accom-plished at two basic levels of abstraction. The lower level uses the existing gate-level simulation and power-estimation tools at a gate-level description of an instruction-processing system. This approach uses low-level models of microcontrollers and is the most accurate. However, such models are either unpractical or impossible. A detailed microcontroller description is unlikely to be available to a programmer, especially if the processor is not an in-house designed Application-Specific Instruction Processor.

A higher-level approach is to estimate the power based on the frequency of execution of each type of instruction or instruction sequence (i.e., the execution profile). The execution profile can be used in different ways. The architectural-level power estimation determines [171] which major microcontroller components will be active in each execution cycle of a program. This method is less precise but much faster than the gate-level estimation. It requires a model of the processor's major components, as well as a model of the components which will be active as a function of instructions being executed.

Another approach is based on the premise that the switching activity on buses (address, data and control) is representative of the switching activity (and the power dissipation) in the entire processor [172]. Modelling of the bus switching activity requires knowledge of the bus architecture of a processor, op-codes for the instruction set and a representative set (or statistics) of input data to a program. It also needs information about the manner in which the code and data are mapped into the address space.

The last but not least approach is referred to as an 'instruction-level power analysis' [173,174]. This approach is based on the instruction-level power model and requires that power costs associated with individual instructions and pairs of consecutive instructions will be characterized empirically for the target microcontroller.

The methodology for modelling power at the software level is based on physical measurements. A detailed description of the internal design of the target processor is not required. Only some understanding of the processor architecture is important in order to choose the appropriate types of instruction sequences to be characterized. This knowledge becomes even more important when a programmer looks for ways ᵇ :ed on power estimation to optimize a program.

Two main requirements for instruction-level power analyses are to characterize the power dissipation associated with each individual instruction (base energy cost) and the influence of the instruction sequence on the energy cost (inter-instruction effects because of circuit states). The most straightforward and precise method is to measure directly the current draw of the processor in the target system as it executes various instruction sequences.

The base cost, which is independent of the prior state of the microcontroller, can be determined by putting several instances of that instruction into an infinite loop. The average power supply current is measured while the loop executes. The loop should be made as long as possible to minimize an estimation error due to the loop overhead (the jump statement at the end of the loop). An ammeter is used to measure the current draw of the processor in the system. Its relative error must not be more than 0.1%. In order to minimize the estimation error for embedded microcontrollers it is expedient to take advantage of the more precise method of measurement for base energy costs.

This method of measurement is based on the fact that embedded microcontrollers for low-power applications have a built-in on-chip ROM for programs. For more accurate measurements of instruction base energy cost, the whole volume of the internal ROM is filled in from zero to the greatest possible address by the same code, corresponding to the chosen instruction. It will be automatically reset to zero by the execution of an identical instruction sequence by the attainment of the maximal address by the program counter. And the execution of the instruction sequence, recorded from zero address in ROM, begins all over again. Such a method of measurement is more accurate. It completely removes

Table 9.2   Code mapping in ROM for traditional (1) and new (2) Method

Address	Code	Instruction
	Method 1	
0000	E9	M1: MOV A, R1
0001	E9	MOV A, R1
...	...	...
4094	01	AJMP M1
4095	00	
	Method 2	
0000	E9	MOV A, R1
0001	E9	MOV A, R1
...	...	...
4095	E9	MOV A, R1

the estimation error due to the loop overhead (the jump statement at the end of the loop) and measures not the average but the real power supply current for the instruction.

Fragments of the instruction sequence for traditional (1) and offered (2) methods of measurement for the base energy cost for the instruction MOV A, R1 are shown in Table 9.2.

The given approach is fair for measurement of the base energy cost for data transfer instructions, arithmetic and logic instructions and the unconditional jump. However, while using any of the methods of measurements, mentioned above, it is necessary to take into account the features inherent to the execution of conditional jumps. During the execution of such instructions, depending if the given condition is taken or not, the base energy costs for the same command are various. Moreover, similar models of microprocessors and microcontrollers need more cycles when executing instructions, to execute a jump when it is taken.

Let us consider some examples of the estimation of base energy costs for the two-byte instructions: JC (jump if carry bit C is '1') and JNC (jump if bit C is not '1'). As a rule, after the initialization of a processor by the signal RESET, the carry bit C is set to '0'. In the first example, all instructions JC are executed consistently, since C = 0 and jump on the <label> does not happen. In this case, the instruction sequence for measurement of the base energy cost is similar, as for the considered instruction MOV A, R1.

Example 1

```
0000 JC < label >

0002 JC < label >

             .


             .


             .

4094 JC < label >
```

Let us investigate the situation when the jump is taken. Following processor initialization it is necessary at first to execute the instruction, which sets C to '1'. It is then possible to write two instructions JC with crossed labels of a control transfer for organizing the continuous cycle:

Example 2

```
0000        CPL  C;     C = 1

0001 M0:  JC  M1;    because of C set in "1", jump  to  M1

0003 M1:  JC  M0;    because of C set in "1", jump  to  M0
```

As it can be seen from Example 2 for accurate estimation of the base energy cost for the JC instruction three instructions are sufficient. However, between these instructions there can be any number of commands, which do not influence the results of the measurement, because the logic of this program fragment makes these instructions non-executable.

Let us consider now the instruction JNC (jump if C is set to '0'). In the first case, if the jump is taken, the test partial program is the similar to Example 2 but without CPL instruction. The exception is the absence of the command for the condition initialization, since by the processor initialization C is already set to '0':

Example 3

```
0000  M0:  JNC  M1

0002  M1:  JNC  M0
```

Another situation arises in the case when the condition is not executed:

Example 4

```
0000 CPL C; C=1

0001 JNC < label >

0003 JNC < label >

     .              .

     .              .

4094 JNC < label >
```

As the condition $C = 0$ is not taken, the jump to the <label> does not happen and the next instruction JNC is consistently executed and so on. An estimation error due to the loop overhead (CPL C instruction at the beginning of the loop) arises. However, this error is less than when using the jump statement at the end of the loop.

Base costs for each instruction are not always adequate for precise software power estimation. Additional instruction sequences are needed in order to take into account the effect of a prior processor state on the power dissipation of each instruction. The circuit state cost associated with each possible pair of consecutive instructions is characterized for the instruction level power analysis by measuring the power supply current while executing an alternating sequence of two instructions in an infinite loop. However, the circuit state effect is not symmetric. The following code sequences are considered as an example:

Example 5

```
ADD  A  @R0        ;A      MOV R3, #E7h   ;B

MOV R3, #E7h     ;B      ADD  A,   @R0    ;A
```

The known technique of measurement for the circuit state overhead does not permit the separation of the cost of an A $\rightarrow$ B transition from a B $\rightarrow$ A transition, since the current measurement is an average over many execution cycles [170]. In this case, it is expedient to take an advantage of the proposed indirect method of measurement for more accurate measurement. This method can be used for microcontrollers and microprocessors, admitting the hardware realization of a step-by-step mode for the instruction execution.

The idea behind the method consists of the following. The instruction sequence consisting of two tested commands is written to the ROM. With the help of an additional T-flip-flop, entered into the microcontroller circuit in a step-by-step mode, the division of the frequency of the ALE signal (the identifier of the machine cycle) into two is realized. It executes two instructions during one 'step'. The maximal voltage V_{max} on the resistor R_0 during the execution of the two test instructions is measured by a peak detector. The maximal power supply current is determined as $I_c = V_{max}/R_0$. For example, if the estimated base energy cost for these instructions is I_{es} and the energy measured by the mentioned above method is I_m, then the circuit state overhead will be given by

$$I_{oh} = I_m - I_{es} \qquad (9.1)$$

This method allows the A $\rightarrow$ B as well as the B $\rightarrow$ A transition to be measured. Moreover, while executing only one instruction per step, as stipulated by the standard usage of the step-by-step mode, and using a peak detector, it is also possible to measure the base energy cost for a single instruction.

The total energy for the partial program is

$$I_{\text{Total}} = \sum_{i=1}^{n} I_{bi} + \sum_{j=1}^{k} I_{ohj}, \qquad (9.2)$$

where I_{bi} is the base energy cost for the ith instruction; I_{ohj} is the circuit state overhead for the jth pair of instructions; n is the number of instructions in the tested partial program; k is the number of tested pairs. The number k is determined by the functioning logic of the tested partial program.

The evaluation stage of worst-case power consumption situations in embedded processors is the next important step for optimization. In other words, it is necessary to verify if a processor meets power constraints. The known technique of this stage includes the following stages [175]:

- Construct a power consumption graph based on an instruction level power model.

- Find the cycle in the graph with the maximum mean power consumption.

An oriented graph $G(V, A)$ is used as a power consumption graph, where $V = \{v_i\}$ is the vertex set; $A = \{\gamma_{ij}\}$ is the set of arches. There are the following model restrictions:

1. The graph has one input V_0 and not more than one output V_k.

2. No more than two arches emanate from each vertex.

3. The number of arches entering the vertex is not limited.

The graph corresponds to the algorithm's decision tree; nodes to the algorithm's instruction. In the traditionally used graph, each node is characterized by the additive elementary index d_i, which is connected with the base energy cost only. However, with the aim of more accurate estimation for the total energy, it is necessary to take into account the above estimated circuit state overheads. For this purpose, it is suggested modifying the graph, having connected each of the arches with the additive elementary index g_j, which characterizes the circuit state overheads. Entered indexes from the set $D = \{d_i, g_j\}$ are shown on the graph (Figure 9.2). After that, the matrix is built and Karp's algorithm is used for the determination of the maximum average weight for a critical cycle, according to the technique [175].

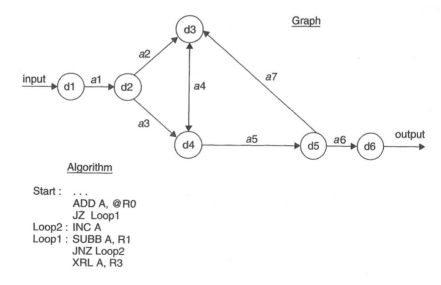

Figure 9.2   Power consumption graph

The graph $G(V, A)$ can be used for the static research of the power consumption for various parts of the algorithm. However, the dynamics of the execution of a partial program is determined by choosing a certain pattern on the graph. This choice arises from the control transfer in logic blocks, which are connected with the casual process of ingress of various vectors of the data $Y^{(0)}$ into a program. It results in the casual choice of patterns on the graph. Thus, the tested partial program is a complex system with a casual structure. From the point of view of power optimization, the task is complicated by the knowledge of how long and often the microcontroller executes this worse cycle. Consequently, to achieve a real power saving the partial program should be optimized similar to the total program. Let us consider possible ways of power or energy minimization for microcontrollers. We assume, that (1) for the solution of the task an adequate algorithm and data structure is used; (2) assembly language is used for source-code programming. This allows opportunities for a programmer alone to operate the hardware, the execution time of a program and, as will be shown further, the power consumption.

9.2.1   Instruction selection and ordering

As there are usually many possible code sequences that accomplish the same task, it should be possible to select a code sequence that minimizes power and energy. It seems obvious, but to obtain an optimum decision is not so simple in the majority of real situations. One such example is choosing between two instructions. Each of them can be used for data moving from the accumulator A to the register R0 (with various side effects): MOV R0, A or XCH A, R0. For the microcontroller family MCS-51 (Intel) each of these instructions is executed for one cycle. Another example is choosing an instruction for the accumulator clearing: XRL A, A; CLR A or MOV A #0h. Multiplication by two can be executed through the arithmetic instruction for multiplication MUL AB as well as through the left shift instruction. In general, the choosing condition can be written as

$$K\{d_i, a_{i-1}\} \longrightarrow \min \tag{9.3}$$

Instruction ordering for low power attempts to minimize the energy associated with the circuit state effect without altering the program behaviour. In reality, adjacent commands can change their location frequently without damage to the logic of the algorithm:

A. DEC   @R1   B.   CPL   C

    CPL   C   DEC   @R1

9.2.2   Code size and speed optimizations

Many code optimization techniques used for speed or code size should be equally applicable to the power or energy optimization. This is a very important aspect, especially for embedded systems. The code optimization used to reduce the size assumes the merging of all identical or functionally similar code sequences in the subroutines. It has been discussed above in Chapter 7, Equations (7.71–7.73).

Low power software system design strongly depends on the application. However, some common principles can be highlighted. The basic idea is to reduce the number of basic steps and clock cycles necessary for the execution of a given task. If this number N of instructions is reduced, the microcontroller can be in the idle or power down state for a longer period, or the basic frequency can be reduced as well as V_{DD}.

All these factors reduce the power consumption significantly. The execution time of a task in a microcontroller is given by:

$$T_{execution} = N \cdot n \cdot \tau_{clc} \qquad (9.4)$$

where N is the number of instructions executed to perform the task; n is the number of cycles or phases per instruction; τ_{clc} is a cycle or a phase duration, i.e. $\tau_{clc} = 1/f_{clc}$.

The code size reduction in embedded systems is directly linked to energy savings. The size of built-in ROM in microcontrollers is from 1 up to 32 kb. The increase of an executive program by 1 byte in comparison with the memory size of programs, requires the use of external additional memory. And this, in turn, results in increasing the power consumption because of the violation of a low-power programming principle–to use only the internal memory. Very often measures on the code size optimization are opposite to measures on the speed optimization. Sometimes, non-obvious aspects of optimization can cause code size reduction as well as an increase in speed (the reduction of the task execution). Consequently, the low-power programming style must use power-saving techniques directly as well as code size and speed optimizations. These techniques also influence the power consumption. The choice of one or other optimizations depends on the particular application and hardware restrictions. Let us investigate some low-power programming 'tricks' in detail that should be useful for the smart sensor design.

The reduction of N can be achieved in different ways. There are some common procedures for the 'local' optimization:

- The reduction of the number of control transfers in a program:

    - at the expense of subroutines' transformations and direct inclusion of their contents in the executive code (a violation of the principles of structural programming for the low-power programming style);

    - at the expense of condition jump instructions transformations of conditional transitions so that the jump is true relatively less often than the condition for its absence;

    - moving of general conditions to the beginning of the fork sequence of jumps;

    - call transformations, followed by the immediate return into the program to unconditional jumps.

- The cycle optimization, including moving of the calculation of non-changeable values outside the cycle; cycle unrolling and merging of the separate cycles, which are executed the same number of times into the uniform cycle (the cycle 'compression').

9.2.3   Jump and call optimizations

The empirical rule for jumps and calls is very simple: to get rid of them where possible. The program should be organized so that it is executed by a direct, consecutive way, with a minimum number of branchpoints. The further increase of productivity is realized, for example, by transformations of the jumps to a common exit point in multiple exits. Subroutines can be transformed to instruction sequences of the main program. In this case, there are no additional time costs for the control transmission and return. If the program requires condition jump instructions, it is expedient to organize it so that the probability of jumps is lower than its absence. For example, the program checks the seventh bit in the accumulator, which can be set to '1' only in rare cases and by specific circumstances. In this case, it is expedient to construct the program so that it may 'rush' through the branchpoint, if the seventh bit of the accumulator is set to '0':

Example 6

```
ANL   A,  R3

JB7   L1       ;  happen to be rare.
```

Let us consider one more case, when different values of some variable initiate different actions and require the multiple matching, followed by condition jump instructions. In this case, it is expedient to move the matching with large probable values closer to the beginning of the program:

Example 7

```
        CJNE R1, #2Ah, Loop1     ;the most probable value

        ...

        JMP Loop3

Loop1:  CJNE R1, #9Eh,  Loop2    ;the less probable value

        ...

        JMP Loop3

Loop2:  CJNE R1, #FFh,  Loop3    ;the least probable value

        ...

Loop3:
```

There are certain ways of optimization, connected with jumps and calls, which require the addition of a certain degree of 'destructiveness' to the program. Each of them means the transformation of calls to jumps. The calls themselves require more time for execution than jumps, since the stack address for return requires more frequent memory access. It violates the next principle of a low-power programming style–to minimize the memory access. The next low-power 'trick' consists of the following. If instructions LCALL or ACALL for the nested subroutine are followed by the instruction RET for return to the top-level program, such a sequence can be transformed to the unconditional jump–JMP:

Example 8

```
Before:                                    After:
-------------------------------------------------
PROG1:     CLR  A                PROG1:     CLR  A

           . . .                            . . .

           ACALL     PROG2       JMP        PROG2

           RET

PROG2:     ORL  A,  @R3          PROG2:     ORL  A,  @R3

           . . .                            . . .

           RET                              RET
```

The effect from such an optimization consists of the following. As far as the instruction address causing the PROG1 is in the stack on the PROG2 input, the procedure PROG2 comes back to the calling program directly. Thereby, superfluous instructions ACALL and RET are eliminated. If the program PROG1 (in the ROM) is followed physically by the program PROG2, it is possible to do without the JMP PROG2 instruction, and the PROG2 can be followed by the PROG1 execution.

9.2.4   Cycle optimization

All existing methods for the cycle optimization can be reduced to a set of empirical rules:

- Never to do inside the cycle, what is possible to do outside of it.

- Everywhere, if possible, get rid of the control transfer inside cycles.

The first rule results from the old truism: 90% of the execution time of a program has 10% of code. And more often these are various cycles. Therefore, the first thing that should be done to reduce the execution time of a program is to find the 'hot' pieces of the program and check in all cycles as potential objects of optimization. The following example demonstrates the use of the first rule:

Example 9

```
Before:                                    After:
--------------------------------------------------
        MOV R4,   # 8h              MOV R4, # 8h

        MOV A,    @R3               MOV A,   @R3

REPEAT: RLC A              REPEAT:  RLC A

        MOV R2,   #E9h             DJNZ R4, REPEAT

        DJNZ R4, REPEAT            MOV R2, #E9h
```

The next optimization allows to the elimination of matching and condition jump instructions. It consists in the cycle merging. The instructions are reorganized to build one cycle from several, which are repeated the same number of times. Results of such optimizations are shown in Example 10:

Example 10

```
Before:                                    After:
--------------------------------------------------
        MOV  A,  @R0               MOV  A, @R0

        MOV  R1, #FFh              MOV  R1, #FFh

RPT1:   RL   A             RPT1:   RL   A

        DJNZ R1, RPT1              INC  R3

        MOV  R2, #FFh             DJNZ R1,   RPT1

RPT2:   INC  R3                    MOV  R2, #FFh

        DJNZ R2,  RPT2
```

Another way to eliminate cycles is not to use small loops, i.e. to remove control code sequences from the cycle, by simply repeating the cycle body a necessary number of times. This gives an especially good result, when the time, necessary for the cycle body execution, is less than the time for cycle control operations:

Example 11

```
Before:                                    After:
--------------------------------------------

        MOV  A,  @R0                       MOV A, @R0

        MOV  R1, #4h                       RR   A

Loop1:  RR   A                             RR   A

        DJNZ R1, Loop1                     RR   A

                                           RR   A
```

In some cases, the method considered in the Example 11 for unrolling loops with the aim of increasing speed, also causes code size reduction. It is true, if the following inequality is correct:

$$V_{j\text{control}} + V_{j\text{cycle}} \geq V_{j\text{cycle}} \cdot k_i, \qquad (9.5)$$

where $V_{j\text{control}}$ is the memory size, necessary for control instructions of the cycle (in Example 11 it is the MOV and DJNZ instructions); $V_{j\text{cycle}}$ is the cycle body size; k_i is the number of cycle executions.

The next 'trick' also reduces the code size and increases the speed of the program execution, by using interrupts. As far as the interrupt vector addresses for microcontrollers are allocated, as a rule, with the interval in some bytes, the small interrupt routine can be completely located within the limits of these bytes. Moreover, if other interrupt vectors are not used, then the interrupt routine can increase in size. For example, in case of one external interrupt and the interrupt from the microcontroller's serial port, in order to allocate the interrupt routine for the external interrupt processing, memory address space from 0003h to 0023h will be accessed. Such allocation for the interrupt routines saves one unconditional jump instruction JMP.

Sometimes measures used for power saving can result in performance deterioration, for example, speed reduction. In order to compare the efficiency of accepted measures, it is expedient to take advantage of the factor for the relative efficiency:

$$K_{\text{eff}} = \frac{P_S}{S_L}, \qquad (9.6)$$

where P_S is the factor describing the power saving; S_L is the factor describing the speed loss. For example, if $P_S = 30$ X and $S_L = 3$ X, then $K_{\text{eff}} = 10$. The higher K_{eff} the better.

9.2.5 Minimizing memory access cost

Memory often represents a large fraction of a system's power budget and there is a clear motivation to minimize the memory power cost. Fortunately, software design techniques that have been used to improve the memory system efficiency can also be helpful in reducing the power or energy dissipation of a program. This is largely due to the memory being both a power and a performance bottleneck [170]. If memory accesses are slow and power hungry, then software techniques that minimize the memory accesses tend to have both a performance and a power benefit. For embedded systems, power minimization techniques related to the memory concentrate on the following objectives:

- Minimize the number of the memory accesses required by an algorithm.

- Minimize the total memory required by an algorithm.

- Make the memory accesses as close as possible to the processor: register first, the internal RAM next, the external RAM last.

We shall go into detail about these items, because the code size minimization and the cycle optimization considered above can substantially influence the number of memory transfers and total storage requirements.

9.2.6 Exploiting low-power features of the hardware

There are a number of current techniques and ideas for future low-power design that require consideration in the software design process. Such low-power techniques include the software control of power management.

Some microcontrollers support varying levels of the software control over power management. Some examples include the Intel families MCS-51/251, MCS-96/196 and Texas Instruments family MSP430, which have a 'power-down' register that masks or enables the clock distribution. Such microcontrollers provide two power reduction modes: 'idle' and 'power down'. The idle mode shuts off the CPU but peripheral devices and RAM are active. In this mode, the power consumption is 15% from normal. The power-down mode halts all operations but saves data in the RAM. In this mode, the current supply is less then 10 μA. Such a technique obviously reduces the activity a_i to zero during the idle state. The exit from these modes is executed by the interrupt or the RESET signal. At the exit from the idle mode by the RESET signal, it is necessary to take into account, that between a reset of the bit in the 'power-down' register and the start of the initialization algorithm, about two machine cycles can take place. This does not exclude the command execution, after the processor switch in the idle mode. Thus, it is undesirable, that the reset of a bit in the 'power-down' register is followed by the MOV Px, src instruction, which distorts the values in I/O ports. The software designer is in a position to determine when this or that regime will be of benefit. But an incorrect decision to power down can hurt the performance and the power dissipation due to the cost of restoring power and the system state.

One of the most suitable microcontrollers for use as an 8051-microcontroller compatible core for microsystem design is D87C51 FA [176]. The chart of dependence of the chip consumption current from clock frequencies is shown in Figure 9.3. In the power-down mode the supply current is 75 μA.

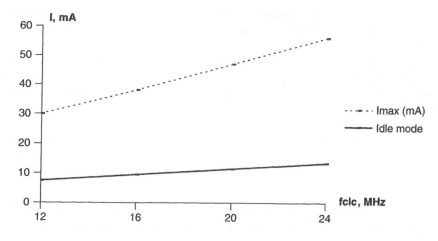

Figure 9.3   Dependence of supply current on clock frequency

9.2.7   Compiler optimization for low power

A variety of optimizations and software design approaches for energy consumption minimization need to be incorporated into automated software development tools. There are two levels of tools: tools that generate a code from an algorithm specification and the compiler level optimization [170]. Such tools exist for DSP applications, however, they do not exist for embedded microcontrollers.

Compiler technology for low power appears to be further along than high-level tools in part because well-understood performance optimizations are adaptable to the energy minimization. Considerable work has been done in compiler optimization for maximum performance, but little consideration has been given to the optimization for low power. Existing compilers allow the user to select between optimizing for the code size and optimizing for the performance; adding a third, user-selectable criterion to cause the compiler to optimize for power efficiency is quite feasible [167]. However, the software power optimization is based on the preliminary software power estimation, which in its turn, is based on physical power measurements. It is one of the deterring factors for low-power compiler design. To estimate all base energy costs and circuit state overheads taking into account its non-symmetric inter-instruction effects for a particular microcontroller's instruction set, it is necessary to carry out the following number of measurements:

$$N = n + 2 \cdot C_n^2, \tag{9.7}$$

where n is the number of microcontroller instructions. However, probably, for the majority of microcontrollers, it is possible to unite all instructions into groups. Within such groups, the base energy cost and circuit state overheads are the same. Then the number of measurements can be reduced by a greater order. The global decision for the problem consists in the standardization and automatization of the base energy cost and circuit state overhead measurements for various microcontrollers and microprocessors. It is expedient that such measurements are made only once, and their results should become widely accessible (for example through the Internet). In practice, to accomplish particular tasks for the 'local' optimization by the end user, it is not obligatory to carry out the total number of measurements. From the whole set of existing instructions, half is used in practice and the optimization task can be successfully resolved on the empirical level.

The use of the modern technique of 'low-power programming style' solves the problem of optimization of a partial program for low-power efficiency in a novel way. Such optimization (basically on three parameters: power–speed–code size) is necessary for the following reasons:

- Methods of modern technologies of programming, for example, structural programming (even if it makes software development for embedded systems easier), frequently reduces the programs' efficiency for the clearness, the reliability and the low cost.

- The wide usage of high-level languages and the imperfection of optimizing compilers also results in decreasing efficiency of the developed software.

The high power saving efficiency must be provided by the:

- correct analysis of those parts of the program, that need the optimization

- automation of processes for software optimization by the application of means for software power analysis and estimation

- usage of means of automation of optimizing transformations for the software (using special optimizing procedures)

- analysis of those programs' parts, which allow optimizing transformations.

According to [177], dynamic performance execution processes for the software (we designate it λ) by the microcontroller, the average time T_λ, for the program execution can be given by the mathematical expectation

$$M[T_\lambda] = \sum_{i=1}^{m} p_i \left(\sum_{j=1}^{l_i} \tau_{ij} \cdot y_{ir} \right) \tag{9.8}$$

where m is the number of parts (linear and elementary cyclic areas of the λ); p_i is the probability of execution for the ith program part; l_i is its length; τ_{ij}, y_{ir} are the duration and frequency of execution of the jth instruction of the ith fragment.

Let some set optimizing procedures $\{PR_\gamma\}$ be given. For the instruction optimization for the low-power programming style the set includes: instruction selection and ordering; jumps, calls and cycles optimization; loop unrolling, and code size minimization (subroutine merging). It is necessary, using this 'hints' set, to reconstruct the program parts (and, maybe, its structure G_λ), to minimize the power consumption and the time for execution, e.g.

$$\sum_{i=1}^{m} p_i \left(\sum_{j=1}^{l_i} \tau_{ij} \cdot y_{ir} \right) \longrightarrow \min \tag{9.9}$$

$$W_p = \varphi(K) = \sum_{k=1}^{r} P_k + \sum_{l=1}^{T} P_c \longrightarrow \min \tag{9.10}$$

by the restriction on the memory size, which after the optimization should not exceed some given size, e.g.

$$\sum_{i=1}^{m}\sum_{j=1}^{l_i} V_{ij} + \sum_{n=1}^{N} V_n \le V_\lambda, \tag{9.11}$$

where V_{ij} is the memory size for storage of the jth operator of the ith partial program; V_n is the size for storage of the nth operand; P_k is the base energy cost for the kth instruction; P_c is the circuit state overhead for tth pair of instructions because of the inter-instruction effect; r is the number of instructions in the partial program being optimized; T is the number of tested pairs. The number T is determined by the functioning logic of the partial program.

By the fixed set $\{PR\}_\gamma$ the task can be reduced to determination of the program parts, which are subjected to optimization. The instruction selection and ordering 'hints' reduce the base energy cost and the circuit state overhead directly. In its turn, the speed optimization reduces the number of basic steps and clock cycles necessary for the execution of a given task. If this number N of instruction is reduced, the microcontroller can be in the idle or power-down state for a longer period, or the basic frequency can be reduced as well as V_{DD}.

As the statistics show, for the majority of programs the significant part of time (up to 90%) falls on the execution of a small (5–10% operators) part of the program (critical area). Therefore, the power consumption can be reduced by optimization of performances of the most frequently executed partial programs. We rationalize fragments on their weights w_i in the power budget, e.g. we construct the sequence

$$\langle \chi_i \rangle :_{wi \ge wj} \ge w_i \longrightarrow \chi_j \longrightarrow \cdots \tag{9.12}$$

In the optimizing set the partial program x_i from the sequence is included (9.12), for which it is true

$$\Omega = \left\{ \chi_i \in X \left| \sum_i w_i \ge 0.9 \right. \right\} \tag{9.13}$$

We enter the set of Boolean variables $H = \{h_i\}$, where $h_i = 1$, if the fragment x_i can be optimized; $h_i = 0$, otherwise.

As a result of optimization of the ith partial program let its execution time, memory size and power consumption change accordingly to sizes Δt_i, ΔV_i and $\Delta W p_i$. In general, the first two values can be positive as well as negative. Then the task (Equations 9.9–9.10) is reduced to the determination of H_0 by which is reached

$$\max \sum_{i \in J_0} \Delta t_i \cdot h_i \tag{9.14}$$

by restrictions

$$\sum \Delta t_i h_i \le V_\lambda - \left(\sum_{i=1}^{m}\sum_{j=1}^{l_i} V_{ij} + \sum_{n=1}^{N} V_n \right) \tag{9.15}$$

$$\forall_i \in J_0 (h_i \in \{0, 1\}) \tag{9.16}$$

The task (9.14–9.16) is the task of linear programming with Boolean variables. Such a task can be solved by the total search for all variants and can be prescribed in the base of optimizing compilers for low power. However, in comparison with the creation of conventional optimizing compilers for the speed and code size, some problems arise. If the application of measures used for the speed and code size optimizations does not infringe the program logic, the reordering, with the aim of reducing the circuit state overhead, can result in such infringement. Let us consider the following example:

a)	is		b)	is not	
DEC @R1	*equivalent*	MUL AB	MOV A,@R1	*equivalent*	MUL AB
MUL AB	*to*	DEC @R1	MUL AB	*to*	MOV A,@R1

In the first pairs (*a*), the instructions are mutually independent and can be changed by locations. By this, the program logic will not be violated. In case (*b*), the result of the instruction execution of the multiplication MUL depends on the number, which will be loaded in the accumulator A by the instruction MOV. Its reordering causes the wrong result.

Compiler design for low power, which takes into account the program logic, represents a the difficult task. In the first stage, it could be limited by the compiler design, admitting the work in the interactive mode. During the compilation, variants of reordering are granted to the programmer from allowable ones are chosen.

It is possible to make the following conclusions:

- Due to low costs of microcontrollers or core processors, software low-power design for various smart sensors is becoming a 'hot' topic.

- Partial techniques of local optimization must be used, if the algorithms and structures of data are adequate.

- The 'tricks' of the low-power programming style help in making software design solutions consistent with the objective of power minimization.

- For embedded systems, methods of software low-power optimization are connected closely with code size and speed optimizations.

- By preliminary estimates, the power saving at the software level, may be from 10 to 35%.

Summary

For embedded microcontrollers in smart sensor design it is expedient to use some optimization at the software level. First of all this means using the set instruction minimization for metering applications and optimal low-power programming. The first measure reduces the chip area, the second saves power consumption up to 35%. Both are very important aspects for smart sensors and improve sensor characteristics practically without additional labour expense and time.

10

SMART SENSOR BUSES AND INTERFACE CIRCUITS

In large measurement and control systems (modern cars, for example), which include hundreds of sensors, communication between a central computer and the smart sensors, which are widely scattered, is difficult. It is probable that the sensors are produced by different manufacturers, and that the sensors' outputs have different formats. In order to design efficient sensor systems the data transfer must ideally be organized in an orderly and reliable fashion. Such systems are called bus systems or networks. In general a bus system consists of a central computer that is connected by a number of wires to a large number of sensors. When a sensor is activated to send information into the central computer, the sensor's address is selected and the sensor will be switched to the digital data line. The central computer can initiate different kinds of tests and recalibrations. Every sensor is connected to the same wires. A dedicated transmission protocol is applied to allow a flexible and an undisturbed data flow.

There are many buses now available and it is still difficult to find the one to suit smart sensors' requirements. We do not intend to describe in detail all existing sensor buses and network protocols in this book. For more detailed information we recommend references [8] and [178]. The mission of this last chapter is to show the most applicable digital bus interfaces for smart sensor applications as well as to answer what is expedient to do with analog sensors.

10.1 Sensor Buses and Network Protocols

A number of different protocols exist, each having its own interface requirements. The requirements stipulate such parameters as headers, the data-word length and type, the bit rate, the cyclic redundancy check, and many others. Table 10.1 shows distinct smart-sensor network protocols.

There are some interface devices available, for example, Motorola, produces versions of its 68HC705 microcontroller that provides a J1850 automotive-network interface and variations of the company's 68HC05 microcontrollers that incorporate computer-automated network interface functions.

All these interfaces are strongly associated with their field of application. A close examination of existing buses shows that these are designed to fit a special set of requirements, for example domestic, industrial and automotive applications, measuring systems, etc. as shown in Table 10.1. Smart sensors are a new kind of application,

Table 10.1   Sensor network protocols

Sensor network protocol	Developer
Automotive	
J-1850, J-1939 (CAN)	SAE
J1567 C^2D	SAE (Chrysler)
J2058 CSC SAE	Chrysler
J2106 Token Slot	SAE (General Motors)
CAN	Robert Bosch GmbH
VAN	ISO
A-Bus	Volkswagen AG
D^2B	Philips
MI-Bus	Motorola
Industrial	
Hart	Rosemount
DeviceNet, Remote I/O	Allen-Bradley
Smart Distributed Systems	Honeywell
SP50 Fieldbus	ISP+World FIP = Fieldbus Foundation
LonTalk/LonWorks	Echelon Corp
Profibus DP/PA	DIN (Germany), Siemens
ASI Bus	ASI Association
InterBus-S	InterBus-S Club, Phoenix
Seriplex	Automated Process Control (API Inc)
SERCOS	VDW (German tool manufacturers assoc)
IPCA	Pitney Bowes Inc
HP-IB (IEEE-488)	Hewlett-Packard
Arcnet	Datapoint
WorldFIP	WorldFIP
Filbus	Gespac
Building and office automation	
BACnet	Building Automation Industry
IBIbus	Intelligent Building Institute
Batibus	Merlin Gerin (France)
EIbus	Germany
Home automation	
Smart House	Smart House LP
CEBus	EIA
I^2C	Philips
University protocols	
Michigan Parallel Standard (MPS)	University of Michigan
Michigan Serial Standard (MSS)	University of Michigan
Integrated Smart-Sensor Bus (IS^2)	Delft University of Technology
Time-Triggered Protocol	University of Wien, Austria

therefore the majority of existing digital bus systems cannot be directly applied while maintaining optimum performance.

The I^2C is a very interesting bus from a smart sensor point of view [8]. The topology is rather simple and it has minimum hardware requirements compared to the D^2B. But whereas the hardware specification allows simple interfacing, the communication protocol is too rigid. This prompted the design of a new *Integrated Smart Sensor* bus, IS^2 similar to the I^2C bus, but with a highly simplified protocol. To keep the complexity of the required electronics to a minimum, only the most elementary functions are implemented in the bus. As in I^2C, the IS^2 bus requires two lines for communication: a clock line and a data line. An important feature of the IS^2 bus is that the data field length is not defined. The transmission can be terminated either by the data master or the data sensor [8].

One of the new industrial examples is the colour digital image sensor MCM20027 from Motorola [179]. This sensor is digitally programmable via the I^2C interface.

Another interesting bus from a smart sensor point of view is the Controller Area Network (CAN), which has recently received a wide distribution. It is an advanced serial communication protocol for automotive networking, which efficiently supports distributed real-time control with a very high level of reliability. Originally developed at the beginning of the 1980s by Bosch to simplify the wiring in automobiles, its use has spread to machine and factory automation products. It is also suitable for industrial applications, building automation, railway vehicles and ships. The CAN provides standardized communication objects for process data, service data, network management, synchronization, time-stamping and emergency messages. It is the basis of several sensor buses, such as DeviceNet (Allen-Bradley), Honeywell's SDS or Can Application Layer (CAL) from CAN in Automation, a group of international users and manufacturers, which comprises over 300 companies. CANOpen is a family of profiles based on CAN, which was developed within the CAN in Automation group. The extensive error detection and correction features of the CAN can easily withstand the harsh physical and electrical environments presented by a car. The SDS was developed by Bosch for networking the majority of the distributed electrical devices in an automobile, initially designed to eliminate the large and expensive wiring harnesses in Mercedes.

Other sensors with the CANopen bus are the pressure transducer of COP series with full-scale accuracy up to 0.15% [180] and the microprocessor-controlled CO_2 sensor based on the infrared light absorption from Madur Electronics (Austria) [181]. The latter is also available with the M-Bus interface. Dynisco Europe GmbH, STW and BDsensors have also proposed some new pressure sensors with the CAN interface [182].

Today many semiconductor manufacturers offer microprocessors with embedded CAN controllers. For example, the DS80C390 from Dallas Semiconductor, SAK 82C900 and SAE 81C90/91 standalone CAN-controllers, the 16-bit microcontroller C167CR/CS, C164CI and C505CA, C515C8051-compatible microcontrollers from Infineon Technologies Corporation, 68HC05/08/12 microcontrollers from Motorola, μPD78070Y and μPD780948 8-bit family microcontrollers and V850E/IA1/SF1 32-bit family microcontrollers from NEC Electronics Inc [183].

A sensor bus interface for use in generic bus-organized sensor-driven process control systems was developed at the University of Michigan. The sensor-bus interface is microprocessor-controlled. There are parallel and serial bus structures. Both are suitable for distributed control systems.

10.2   Sensor Interface Circuits

10.2.1   Universal transducer interface (UTI)

A common disadvantage of many digital interfaces is that many analog sensors cannot be interfaced in a low-cost way. In order to eliminate this disadvantage, the Universal Transducer Interface (UTI) circuit for different kinds of analog sensors was designed in the Electronics Research Laboratory, Delft University of Technology [184] and is manufactured by Smartec [185]. It is a complete analog front end for low frequency measurement applications, based on a period-modulated oscillator. Sensing elements can be directly connected to the UTI without extra electronics. Only a single reference element, of the same kind as the sensor, is required. The UTI is intended for the low-cost market, for systems where this interface provides an intermediate function between low-cost sensor elements on the one hand and the microcontroller on the other. This interface, which is directly read out by a microcontroller, services the following sensor elements: Pt resistors, thermistors, potentiometers resistors, capacitors, resistive bridges. With some extra electronic circuitry, the UTI can be used to measure voltage (two ICs MAX4560) and current, which makes them suitable for thermopiles, thermocouples and other types of voltage- or current-output analog sensors. The UTI converts low-level signals from an analog sensor to a period-modulated (duty-cycle) microcontroller-compatible time domain signal [186]. When the sensor signal is converted to the time domain, using a period modulator in the UTI, then the micro-controllers do not require a built-in ADC. So, if the UTI is connected to an Intel 87C51FB microcontroller, for example, it is possible to measure the period with a 0.33 μs resolution.

The signal conversion is carried out according to the linear law:

$$M_i = k \cdot S_i + M_{\text{off}}, \tag{10.1}$$

where S_i is the analog output sensor's signal, k and M_{off} are measuring converter parameters that directly influence conversion error. In order to achieve high accuracy the UTI operates in auto-calibration, which is based on a three-phase differential method of measurement. The given method allows removing the error caused by the above parameters. The essence of this method of measurement consists of the measurement of three signals: $S_1 = 0$, $S_2 = S_{\text{ref}}$ and $S_3 = S_x$ (zero, reference and measurand) during one cycle:

$$M_{\text{off}} = M_{\text{off}},$$
$$M_{\text{ref}} = k \cdot S_{\text{ref}} + M_{\text{off}}, \tag{10.2}$$
$$M_x = k \cdot S_x + M_{\text{off}}.$$

The output signal of the UTI has three informative components and is shown in Figure 10.1.

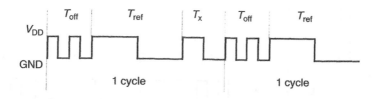

Figure 10.1 Period-modulated output signal of UTI for 3-phase mode

During the first phase T_{off}, the offset of the complete system is measured. During the second phase T_{ref}, the reference signal is measured and during the last phase T_x, the signal itself is measured. The duration of each phase is proportional to the signal that is measured during that phase. The result is the ratio:

$$M = \frac{T_x - T_{off}}{T_{ref} - T_{off}} = \frac{C_x}{C_{ref}} \text{ or } = \frac{R_x}{R_{ref}} \text{ or } = \frac{V_x}{V_{ref}} \tag{10.3}$$

There are 16 different modes with 3–5 phases within one cycle [187]. The connection of capacitors (up to 12 pF, mode 0, 1, 2) to the UTI and time diagram for this mode are shown in Figure 10.2. It is possible to measure multiple capacitances as well as capacitances from 300 pF to 2 pF. Possible applications are liquid level sensors, humidity, position, rotation, movement, displacement sensors. The result can be calculated as follows:

$$M_i = \frac{T_{xi} - T_{off}}{T_{ref} - T_{off}} = \frac{C_{xi}}{C_{ref}}. \tag{10.4}$$

The connection of the thermistor Pt100 (R_x) to the UTI and the time diagram for the modes 5, 6, 7 and 8 is shown in Figure 10.3.

The result can be calculated according to the formula:

$$M = \frac{T_x - T_{off}}{T_{ref} - T_{off}} = \frac{R_x}{R_{ref}} \tag{10.5}$$

The connection of the resistive bridge to the UTI and the time diagram for modes 9 and 10 is shown in Figure 10.4.

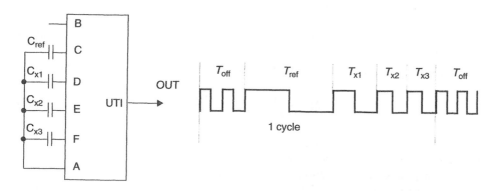

Figure 10.2 Measurement circuit for small capacitance and output signal of UTI

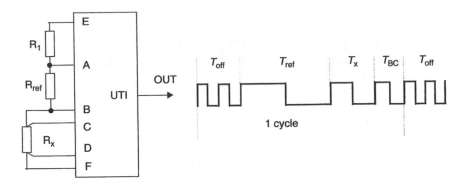

Figure 10.3   Measurement circuit for thermistor and output signal of UTI

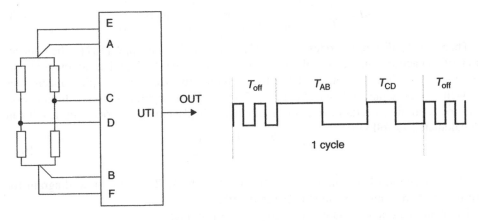

Figure 10.4   Measurement circuit for resistive bridge and output signal of UTI

Possible applications are pressure sensors and accelerometers. The measuring result can be calculated as follows:

$$M = \frac{1}{32} \cdot \frac{T_{CD} - T_{off}}{T_{AB} - T_{off}} = \frac{V_{CD}}{V_{AB}} \tag{10.6}$$

The connection of potentiometers to the UTI and the time diagram for this mode is shown in Figure 10.5.

The measuring result can be calculated according to the following equation:

$$M_i = \frac{T_{xi} - T_{off}}{T_{EF} - T_{off}} = \frac{R_{xi}}{R_{pi}} \tag{10.7}$$

For the measurement of multiple sensing elements it is possible to use one UTI combined with a multiplexer as shown in Figure 10.6:

Using the power down function of the UTI it is possible to build up a multiple-channel measurement system, because the output impedance of the UTI is very high when the power down is low (PD = 0). The interfacing multiple UTIs for the

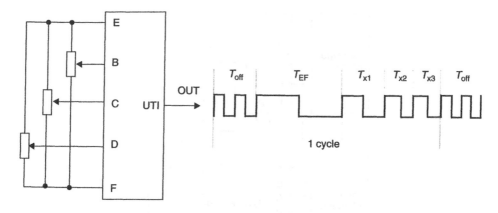

Figure 10.5   Measurement circuit for potentiometer measurement and output signal of UTI

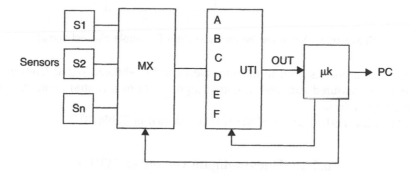

Figure 10.6   Multiple sensing elements measurement

microcontroller is shown in Figure 10.7. To measure a certain channel, the UTI is set to '1' while others must be set to '0'.

The output signal of the UTI can be digitized by counting the number of internal clock cycles in each phase. This sampling introduces a quantization error. The relative standard deviation can be calculated by the following formula:

$$\sigma_q = \frac{1}{\sqrt{6}} \cdot \frac{t_s}{T_{phase}}, \tag{10.8}$$

where t_s is the sampling time and T_{phase} is the phase duration. When the sampling time is 1 µs and the offset frequency is 50 kHz, the standard deviation of the offset phase is 12.5 bits in the fast mode and 15.5 bits in the slow mode. Further improvement of the resolution can be obtained by averaging over several values of the measurand.

Typically, the linearity of the UTI has values between 11 bits and 13 bits, depending on the mode. The UTI is ideal for use in smart microcontroller-based systems. One output-data wire reduces the number of interconnected lines and reduces the number of opto-couples required in isolated systems.

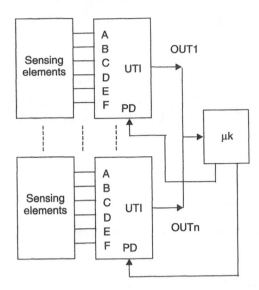

Figure 10.7   Setup for the measurement of multiple channel signal

The microcontroller is used to measure the period-modulated signal from the UTI, to process the measured data and to output digital data to a central computer via the communication interface.

Main technical performances of the UTI are shown in Table 10.2.

10.2.2   Time-to-digital converter (TDC)

Another interesting IC used with quasi-digital smart sensors is the Time-to-Digital Converter (TDC) from Acam-Messelectronic GmbH [188]. It can be used wherever physical quantities need to be digitized for purposes of data processing, with applications including:

Table 10.2   Technical performances of UTI

Parameter	Range
Capacitive sensors, pF	0–2, 0–12, up to 300
Platinum resistor	Pt100, Pt1000
Thermistor, kΩ	1–25
Resistive bridges with max imbalance ±4% or ±0.25%	250 Ω–10 kΩ
Potentiometers, kΩ	1–50
Resolution, bits	14
Linearity, bits	11–13
Measurement time, ms	10–100
Suppression of interference, Hz	50/60
Temperature range, °C	−40− + 85
Power supply, V	2.9–5.5
Current consumption, mA	<2.5

- ultrasonic-based flow and density measurements

- temperature measurements (Pt100, Pt500)

- nuclear and high-energy physics

- laser distance measurement

- ultrasonic position feedback devices

- capacitance and resistance measurement

- frequency and phase measurement (in a few ranges).

By using TDCs the measurement is transferred into time domain signals, making it possible to convert the electronics into a digital single-chip solution.

There is the 2-channel TDC-GP1 and the 8-channel TDC-F1. Some main features of the TDC-GP1 are:

- two measuring channels with a resolution of approx. 250 ps

- two measuring ranges: 2 ns–7.6 μs and 60 ns–200 ms

- four ports to measure capacities, coils or resistors with 16-bit precision and up to 20 000 measurements per second

- an internal ALU for calibration of the measurement results. A 24-bit multiplication unit enables the results to be scaled

- low power consumption (10 μA); full battery operation is possible

- 8-bit processor interface

- ranges for R, L and C measurements: 100 Ω–>1 MΩ, 10 μH–>10 H, 10 pF–>1 mF accordingly.

The key features of the TDC-F1 are:

- eight channels with approx. 120 ps resolution

- optional four channels with approx. 60 ps resolution

- optional 32-bit channels with approx. 5.7 ns resolution

- measuring range of approx. 7.6 μs

- 8-bit I/O interface.

It cannot measure R, L and C parameters or meet the demands of experiments in high-energy physics.

Summary

One of the triggers for the growth of smart sensors is the forthcoming sensor bus standard. The standard will spawn both a wide range of smart-sensor ICs and a generation of sensor-to-network interface chips.

Due to available industrial interface circuits, it is possible to convert analog sensors' signals to the quasi-digital domain easily. The following approach allows an important smart sensor feature: self-adaptation. The interfacing circuits considered above are able to control accuracy and speed using suitable software.

FUTURE DIRECTIONS

Two factors promise to trigger the explosive growth in smart-sensor ICs in the near future. The first is the rapid advances that IC manufacturers are making in microelectromechanical systems (MEMS). Micromachining, both bulk (using deep anisotropic etching) and surface (using thin-film surface layers) are current MEMS processes that many sensor ICs use.

The second growth trigger for smart sensors is the sensor bus standard. The standard will spawn both a wide range of smart-sensor ICs and a generation of sensor-to-network interface chips.

In this market, many companies and countries are in strong competition and as a result there is severe market saturation. The only way to survive in this market is to offer new functions for the same price or to drastically improve the price/performance ratio. One possible way to achieve this is to use novel advanced processing methods and algorithms with self-adopting capabilities.

The degree of integration of microelectronics on a chip can be further increased. Of course, the implementation of the microcontroller in one chip together with the sensing element and the signal-conditioning circuitry is an elegant and preferable engineering solution in the creation of modern smart silicon sensors. However, the combination of monolithic and hybrid integration with advanced processing and conversional methods in many cases is able to achieve magnificent technical and metrological performances for a shorter time-to-market without additional expenditure for expensive CAD tools and a long design process. For the implementation of smart sensors with hybrid-integrated processing electronics, hardware minimization is a necessary condition to reach a reasonable price and high reliability. It is especially true for non-silicon generic technologies like piezoelectric materials, polymers, metal oxides, thick- and thin film materials, optical glass fibres, etc.

The task of creating different smart digital and quasi-digital sensors for various physical and chemical, electric and non-electric quantities is one of the most important and urgent tasks of modern measurement technology. An increase in the accuracy of frequency–time-domain sensors as well as increasing distribution of multiparameters smart sensors are expected in the future.

REFERENCES

[1] *MST Benchmarking Mission to the USA. Report of a European Visit to Microsystems Research and Production Sites in the USA*, 13–25, November 1997, Nexus, 1998.

[2] Huijising J.H., Riedijk F.R., van der Horn G., Developments in integrated smart sensors, *Sensors and Actuators (A: Physical)*, 43, 1994, 276–288.

[3] Novitskiy P.V. Frequency sensors design problem for all electrical and non-electrical quantities, *Measurement Technology*, 4, 1961, 16–21 (In Russian).

[4] Novitskiy P.V., Knorring V.G., Gutnikov V.S., *Digital Measuring Instruments with Frequency Sensors*, Energia, Leningrad, 1970 (in Russian).

[5] Sensor markets 2008, *Worldwide Analyses and Forecasts for the Sensor Markets until 2008*, Press Release, Intechno Consulting, Basle, Switzerland, 1999.

[6] Middelhoek S., Celebration of the Tenth Transducers Conference: The past, present and future of transducer research and development, *Sensors and Actuators (A: Physical)*, **82**, 2000, 2–23.

[7] *Measurement and Automation Catalogue*, National Instruments, USA, 2000.

[8] Rutka M.J., *Integrated Sensor Bus*, Ph.D. Thesis, Delft University of Technology, 1994.

[9] Meijer Gerard C.M., Concepts and focus for intelligent sensor systems, *Sensors and Actuators (A: Physical)*, **41–42**, 1994, 183–191.

[10] Wolffenbuttel R.F., Fabrication compatibility of integrated silicon smart physical sensors, *Sensors and Actuators (A: Physical)*, **41–42**, 1994, 11–28.

[11] Bartek M., *Selective Epitaxial Growth for Smart Silicon Sensor Application*, Ph.D. Thesis, Delft University of Technology, 1995.

[12] Kirianaki N.V., Yurish S.Y., Shpak N.O., Smart sensors with frequency output: State-of-the-art and future development. In *Proceedings of IFAC Workshop on Programmable Devices and Systems (PDS'2000)*, V. Srovnal and K. Vlcek Ostrava (eds), Czech Republic, 8–9 February, 2000, 37–42.

[13] Bianchi R.A., Vinci Dos Santos F., Karam J.M., Courtois B., Pressecq F., Sifflet S., CMOS-compatible smart temperature sensors, *Microelectronics Journal*, **29**, 1998, 627–636.

[14] Riedijk F.R., Huijsing J.H., An integrated absolute temperature sensor with sigma-delta A/D conversion, *Sensors and Actuators (A: Physical)*, **A24**, 1992, 249–256.

[15] Krummenacher P., Smart temperature sensor in CMOS technology, *Sensors and Actuators*, **A21–A23**, 1990, 636–638.

[16] Meijer G.C.M., A three-terminal integrated temperature transducer with micro-computer interfacing, *Sensors and Actuators*, **18**, 1989, 195–206.

[17] Meijer G.C.M., An accurate biomedical temperature transducer with on-chip microcomputer interfacing, *IEEE Journal of Solid State Circuits*, 23(6), 1988, 1405–1410.

[18] Székeley V., Márta Cs., Kohári Zs., Rencz M., New temperature sensors for DfTT applications. In *Proceedings on 2nd THERMINIC Workshop*, 25–27 September, 1996, Budapest, Hungary, 49–55.

[19] Székeley V., Rencz M., Thermal monitoring of self-checking systems, *Journal of Electronic Testing: Theory and Application*, **12**, 1998, 81–92.

[20] Székeley V., Rencz M., Török S., Márta Cs., Lipták-Fegó L., CMOS tempera-ture sensors and built-in test circuitry for thermal testing of ICs, *Sensors and Actuators*, **A71**, 1998, 10–18.

[21] Székeley V., Thermal testing and control by means of built-in temperature sensors, *Electronics Cooling*, 4(3), September, 1998, 36–39.

[22] http://www.micred.com.

[23] Bleeker H., van den Eijnden P., de Jong F., *Boundary-Scan Test*, Kluwer Academic Publishers, Dordrecht, 1993.

[24] SMT 160-30 Temperature Sensor, *Specification Sheet*, Smartec, 1998.

[25] http://www.seabird.com/products/spec_sheets/3Fdata.htm.

[26] http://products.analog.com/products/info.asp?product=TMP03.

[27] http://products.analog.com/products/info.asp?product=TMP04.

[28] http://products.analog.com/products/info.asp?product=AD7817.

[29] http://products.analog.com/products/info.asp?product=AD7818.

[30] http://www.analog.com/bulletins/tsis/.

[31] http://www.dalsemi.com/products/thermal/index.html.

[32] http://www.national.com/catalog/AnalogTempSensors.html.

[33] http://www.gfsignet.com/products/as2350.html.

[34] Gieles A.C.M., Integrated miniature pressure sensor, *Dutch patent application* No. 6817089, 1968.

[35] Blaser E.M., Ko W.H., Yon E.T., A miniature digital pressure transducer. In *Proceedings of 24th Annual Conference on Engineering. in Medicine and Biology*, Las Vegas, Nevada, USA, November, 1971, 211.

[36] Makdessian A., Parsons M., DSSP-based pressure sensors, *Sensors*, January, 2001, 18(1) (http://www.sensorsmag.com/articles/0101/38/index.htm).

[37] Czarnocki W.S., Schuster J.P. The evolution of automotive pressure sensors, *Sensors*, May, 1999, 16(5) (http://www.sensorsmag.com/articles/0599/0599_p52/main.shtml).

[38] Gadzhiev N.D., Garibov M.A., Kasimov M.C., Leschinskiy Y.B., Pressure transducer with frequency output, *Measuring Instruments and Control Systems*, 2, 1979, 27–28 (in Russian).

[39] Maleiko L.V., Malov V.V., Rudenko A.P., Yakovlev I.V., et al. Piezoresonant pressure transducers, *Measuring Instruments and Control Systems*, 9, 1984, 19–21 (in Russian).

[40] Beshliu V.S., Kantser V.G., Beldiman L.N., Beshliu V.V., Coban R.A., Integral gauge pressure sensor with frequency output signal. In *Proceedings of*

International Semiconductor Conference (CAS'99), Sinaia, Romania, 5–9 October, 1999, 2, 491–494.

[41] http://xducer.com/SP550.htm.

[42] http://www.grc.co.uk/grc/sub%20pages/hardware/qpg.htm.

[43] Paros J.M., Fiber-optic resonator pressure transducers, *Measurements and Control*, 1992, 144–148.

[44] Busse D.W., Wearn R.B., Intelligent digital pressure transmitters for aerospace applications (http://www.paroscientific.com/intdigaeroapp.htm).

[45] http://www.pressure-systems.com.

[46] http://www.pressure.com/uk/products/.

[47] Bowling S., Richey R., Two approaches to measuring acceleration, *Sensors*, February, 2000, 17(2) (http://www.sensorsmag.com/articles/0200/41/main.shtml).

[48] http://www.riekerinc.com/Accelerometers.htm.

[49] http://www.analog.com/industry/iMEMS/products.

[50] Kitchin C., Acceleration-to-frequency circuits, *Accelerometer News*, Products & Datasheets Magazines, Analog Devices, Issue 4, 1999.

[51] Podzarenko V.A., System design of tachometers, in *Redundant Numeration Systems, Data Processing and System Designing in Engineering of Informatic Conversion*, Vyshcha Shkola, Kiev, 1990, 5–60 (in Russian).

[52] Deynega V. P., Kirianaki N. V., Yurish S. Y., Microcontrollers compatible smart sensor of rotation parameters with frequency output. In *Proceedings of 21st European Solid State Circuits Conference (ESSCIRC'95)*, 19–21 September, Lille, France, 1995, 346–349.

[53] Deynega V.P., Kirianaki N.V., Yurish S.Y., Intelligent sensor microsystem with microcontroller core for rotating speed measurements. In *Proceedings of European Microelectronics Application Conference, Academic Session (EMAC'97)*, 28–30 May, 1997, Barcelona, Spain, 112–115.

[54] Yurish S.Y., Towards integration of intelligent sensors and virtual instruments for measuring and control applications. In *Proceedings of 3rd IFAC Symposium on Intelligent Components and Instruments for Control Applications*, 9–11 June, 1997, Annecy, France, 321–325.

[55] Digital Inductive Position Sensor 1200, Product description, CEMS, Switzerland, 2000.

[56] Intelligent Opto Sensor, *Data Book SOYDE02B*, Texas Instruments, 1996.

[57] Graaf G.de, R.F. Wolffenbuttel, Light-to-frequency converter using integrated mode photodiodes. In *Proceedings of IMTC'96*, 4–6 June, 1996, Brussels, Belgium, 1072–1075.

[58] Graaf G.de, Riedijk F., Wolffenbuttel R.F., Colour sensor system with a frequency output and an ISS or 12 C bus interface. In *Proceedings of Eurosensors X Conference*, 8–10 September, 1996, Leuven, Belgium, 881–884.

[59] http://www.humirel.com/products.html.

[60] *Sensor Technology, E + E Elektronik Calalogue*, Edition' 99.

[61] Hatfield J.V., Daniels A.R., Snowden D., Persaud K.C., Payne P.A., Development of a hand held electronic nose (H2EN). In *Proceedings of EUROSENSORS XIII*, 12–15 September, 1999, The Hague, The Netherlands, 215–218.

[62] Staples E.J., A new electronic nose, *Sensors*, May, 1999, 16(5) (http://www.sensorsmag.com/articles/0599/0599_p33/main.shtml).

[63] Hök B., Blückert A., Löfving J., Acoustic gas sensor with ppm resolution. In *Proceedings of EUROSENSORS XIII*, 12–15 September, 1999, The Hague, The Netherlands, 631–634.

[64] Akimitsu Ishihara, Shukuji Asakura, A possible motion sensor with frequency output. In *Proceedings of the 18th Chemical Sensor Symposium*, 3–5 April, 1994, Tohoku University, Japan, 10, Supplement A.

[65] Royer G.H., A switching transistor D.C. to A.C. converter having an output frequency proportional to D.C. input voltage, *Transaction AIEE*, July, 1955, 74(I), 322–324.

[66] Van Allen, A variable frequency magnetic-coupled multivibrator, *Transaction AIEE*, July, 1955, 7(I), 356–361.

[67] Tetif L.N., Pulse-width modulator, *Measuring Instruments and Control Systems*, 1967, 4, 46 (in Russian).

[68] Starobinskiy N.M., Kapitonova L.M., Measuring magnetic-transistor voltage-to-frequency converter, *Measuring Instruments and Control Systems*, 1972, 1, 43–44, 46 (in Russian).

[69] Liberson K.Sh., Voltage-to-Frequency Converter, *Patent 493914*, USSR, 1974.

[70] Shvetskiy B.I., Yuzevich Y.V., Taranov G.V., Linear voltage-to-frequency converter, *Control Instrumentation*, Issue 4, 1968, 24–29 (in Russian).

[71] Svirtschev V.A., Taranov G.V., Slave Multivibrator, *Patent 245174*, USSR, 1969.

[72] Shvetskiy B.I., Kirianaki N.V., Taranov G.V., Multichannel pulse-code remote control system for sensors with uniform frequency parameter. In *Information Acquisition and Transmission*, AN USSR, Kiev, 1965, 134–143 (in Russian).

[73] Taranov G.V, Frukht E.S., Field experience of digital remote control system with frequency sensors, *Electric Power Stations*, 1967, 2, 68–73 (in Russian).

[74] Aleksenko A.G., Kolombet E.A., Starodub G.I., *Applications of Precision Analog ICs*, Radio i Sviyaz, 1985 (in Russian).

[75] Gutnikov V.S., *Integrated Electronics in Measuring Devices*, Energoatomizdat, Leningrad, 1988 (in Russian).

[76] Ermolov R.S., Ivashev R.A., Kolesnik V.K., Moroz G.V., *Measuring Devices for Machines and Mechanisms Diagnostics*, Energiya, Leningrad, 1979 (in Russian).

[77] Schagin A., Full-range voltage-to-frequency converter, *Radio*, No. 10, 1987, 31–33 (in Russian).

[78] Kayim R., Analog-to-digital converter with balanced charge and discharge, *Electronics,* 46(11), 1973, 37 (in Russian).

[79] Suetin V., Voltage-to-frequency converter, *Radio*, No. 2, 1984, 43–44 (in Russian).

[80] Patent No. 436439 (USSR), Voltage-to-Frequency Converters, 1974.

[81] Kirianaki N.V., *Remote Metering, vol. III, Balanced Systems of Intensity (Design Principles and Calculations)*, Lvov, LPI, 1979 (in Russian).

[82] Nowinski M., New architectures of integrating analog-to-digital converters. In *Proceedings of PDS'96 International Conference*, Ostrava, Czech Republic, 26–28 November, 1996, 176–184.

[83] Svistunov B.L., Frolov I.N., Sensing device for capacitance sensor, *Measuring Instruments and Control Systems*, No. 12, 1987, 21 (in Russian).

[84] Ménini Ph., Dondon Ph., Blasquez G., Pons P., Modelling of a capacitance-to-period converter dedicated to capacitive sensors. In *Proceedings of 13th European Conference on Solid-State Transducers*, The Hague, The Netherlands, 12–15 September, 1999, 549–552.

[85] Chatzandroulis S., Tsoukalas D., Neukomm P.A., A passive telemetry system with a capacitive silicon sensor suitable for blood pressure measurements. In *Proceedings of Transducers' 99*, Sendai, Japan, 7–10 June, 1999, 1038–1041.

[86] Chatzandroulis S., Tsoukalas D., Neukomm P.A., A pressure measuring system using passive telemetry and silicon capacitive sensor. In *Proceedings of 13th European Conference on Solid-State Transducers*, The Hague, The Netherlands, 12–15 September, 1999, 913–916.

[87] Gayduchok R.M., Kirianaki N.V., Tsukornik G.V., Brych V.G., Tsukornik S.V., Device of Measuring Channelling for Periodometers, *Patent No. 1129541*, USSR, 1984.

[88] Kirianaki N.V., *Non-linear Frequency and Phase Selectors*, Lviv, Vysha Shkola, 1987 (in Russian).

[89] Chizhov V.V., Method of Channeling for Multichannel Frequency-to-code Converters, *Patent 530261*, USSR, 1976.

[90] Voronin N.N., Morozov A.P., Kuznetsov Yu.D., Usov G.A., Device for Frequency Measurement, *Patent 312379*, USSR, 1971.

[91] Kirianaki N.V., Yurish S.Y., Shpak N.O., Noise resistant method of signal transmission without enhancement of code distance. In *Proceedings of the Krajowe Sympozjum Telekomunikacji*, 8–10 September, 1999, Bydgoszcz, Poland, 291–296.

[92] Kirianaki N.V., Mokrenko P.V., Leskiv T.N., Samotiy V.V., Yurish S.Y., *Questions of Theory and Designing of Transmitting Equipment for Telemechanical Systems*, Kiev, OMK VO, 1991 (in Russian).

[93] Kirianaki N.V., *Methods for Digital Data Transmission by Cable Lines with Increased Noise Stability and Speed*, Vestnik LPI, Lviv, No. 267, 1972, 61–67 (in Ukrainian).

[94] Clark G.C., Cain J.B., *Error-Correction Coding for Digital Communications*, Plenum Press, London, second edition, 1982.

[95] Kirianaki N.V., Dudykevych V.B., *Methods and Devices of Digital Measurement of Low and Infralow Frequencies*, Vyshcha Shkola, Lviv, 1975 (in Russian).

[96] Kirianaki N.V., Gayduchok R.M., *Digital Measurements of Frequency-Time Parameters of Signals,* Vyshcha Shkola, Lviv, 1978 (in Russian).

[97] Haward, A.K., Counters and timers, *Electronic Engineering,* 1979, **51**(630), 61–70.

[98] *BurrBrown Applications Handbook*, USA, 1994, 409–412.

[99] Kasatkin A.S., *Automatic Processing of Signals of Frequency Sensors*, Energiya, Moscow, 1966 (in Russian).

[100] Yermolov R.S., *Digital Frequency Counters*, Energija, Moscow, 1973 (in Russian).

[101] Bage R., Time and frequency counting techniques, *Electronic Engineering,* 1979, **51**(630), 47–58.

[102] Patent 4224568, Frequency to digital converter, USA, 1980.

[103] Shliandin V.M., *Digital Electrical Measuring Instruments*, Energija, Moscow, 1972 (in Russian).

[104] Ornatsky P.P., *Automatic Measurements and Measuring Instruments (Analogue and Digital)*, Vyshcha shkola, Kiev, 1986 (in Russian).

[105] Shvetskiy B.I., *Electronic Digital Measuring Instruments,* Tekhnika, Kiev, 1991 (in Russian).

[106] Kudriavtsev, V.B., *Precision Frequency Converters for Automated Monitoring and Control Systems*, Energiya, Moscow, 1974 (in Russian).

[107] Chmykh M.K., Weight method of accuracy enhancement and noise stability of digital frequency counters, *Avtometriya*, 1979, **4**, 135–137 (in Russian).

[108] Konstantinov A.I., *Time*, Publishing House of Standards, Moscow, 1971(in Russian).

[109] Tyrsa V.E., Error of measurement of periodic time intervals at separate quantization of stop impulses and time intervals, *Measuring Technique*, 1988, **2**, 47–49 (in Russian).

[110] Ohmae T., Matsuda T., Kamiyama K., Tachikawa M., A microprocessor-controlled high-accuracy wide-range speed regulator for motor drives, *IEEE Transactions on Industrial Electronics,* IE-29(3), August, 1982, 207–211.

[111] Bonert R., Design of a high performance digital tachometer with microcontroller, *IEEE Transactions on Instrumentation and Measurement,* 38(6), December, 1989, 1104–1108.

[112] Prokin M., Double buffered wide-range frequency measurement method for digital tachometers, *IEEE Transactions on Instrumentation and Measurement,* 40(3), June, 1991, 606–610.

[113] Prokin M., DMA transfer method for wide-range speed and frequency measurement, *IEEE Transactions on Instrumentation and Measurement,* 42(4), August, 1993, 842–846.

[114] Yurish S.Y., Kirianaki N.V., Shpak N.O. Accurate method of frequency-time measurement with non-redundant reference frequency. In *Proceedings of the Conference on Actual Problems of Measuring Technique (Measurement-98)*, 7–10 September, 1998, Kyiv, Ukraine, 78–79.

[115] Kirianaki N.V., Yurish S.Y., Shpak N.O., Methods of dependent count for frequency measurements, *Measurement*, 29(1), January, 2001, 31–50.

[116] Patent 788018 (USSR), Method of Measurement of Frequency and Period of Harmonic Signal and Device for its Realization, Kirianaki N.V., Berezyuk B.M., 1980 (in Russian).

[117] Mirskiy G.Y., *Electronic Measurements*,Radio i Svyaz', Moscow, 1986 (in Russian).

[118] Yurish S.Y., Program-oriented Methods and Measuring Instruments for Frequency-Time Parameters of Electric Signals, *Ph.D. Thesis*, State University Lviv Polytechnic, 1997 (in Ukrainian).

[119] *Strategic DSP Platform Product Guide*, Texas Instruments, USA, Issue 3, 1999.

[120] Patent 1352390 (USSR), Method for Frequency Determination, Kuzmenkov V.Yu., 1987.

[121] Patent 883776 (USSR), Method of Measurement of the Ratio of Two Frequencies of Signals and Device for its Realization, Berezyuk B.M., Kirianaki N.V., 1981 (in Russian).

[122] Kirianaki N.V., Yurish S.Y., Shpak N.O. Precise methods for measurement of frequency ratio. In *Proceedings of the 2nd International Conference on Measurement (Measurement'99)*, 26–29 April, 1999, Smolenice, Slovak Republic, 312–315.

[123] Kirianaki N.V., Yurish S.Y., Program methods of measurements for microcontrollers based measuring instruments. In *Proceedings of the International Conference Programmable Devices and Systems (PDS-95)*, Gliwice, Poland, 1995, 103–110.

[124] Kirianaki N.V., Yurish S.Y., Frequency to code converters based on method of depending count. In *Proceedings of the 14th IMEKO World Congress, Vol. 4B, Topic 4, 2nd International Workshop on ADC Modelling and Testing*, Tampere, Finland, 1997, 276–281.

[125] Kirianaki N.V., Kobylianskiy A.L., Golzgauzer R.V., Hardware-software method of frequency measurement, *Metrology*, 1991, **1**, 32–44 (in Russian).

[126] Figueras J., Low power circuit and logic level design: modeling. In *Low Power Design in Deep Submicron Electronics*, edited by W. Nebel and J. Mermet, Kluwer Academic Publishers, NATO ASI Series, Vol. 337, London, 1997, 81–103.

[127] Calculation of TMS320C2xx Power Dissipation, *Texas Instruments Application Report*, Digital Signal Processing Solutions, 1996.

[128] MSP430 Family, Architecture Guide and Module Library, *Texas Instruments Data Book*, Mixed Signal Products, 1996.

[129] MSP430 Family, Metering Application Report, *Texas Instruments Data Book*, Version 3.0, April, 1997.

[130] Patent 1797714 (Russia), Device for Frequency and Period Measurement of Harmonic Signal, Vinogor L.A., Kirianaki N.V., Russia, 1993 (in Russian).

[131] TMS320C5x DSK, *Application Guide*, Texas Instruments, 1997.

[132] Kirianaki N.V., *Adding and Transmitting Devices for Remote Control Systems with Code Signals*, Lvov, LPI, 1983 (in Russian).

[133] Hrynkiewicz E., *Digital Frequency Multipliers of Square Wave*, Silezian Technical University, Zeszyty Naukowe, No.1148, Gliwice, 1992 (in Polish).

[134] Gutnikov V.S., *Measurement Signal Filtering*, Leningrad, Energoatomizdat, 1990.

[135] Yurish S.Y., Dudykevych V. B., Mokrenko P. V., Program-simulated methods of frequency-time measurements for virtual measuring systems. In *Proceedings of 6th International Conference on Industrial Metrology (CIMI'95)*, 25–27 October, 1995, Zaragoza, Spain, 249–256.

[136] Yurish S.Y., Design methodology for reusable software components of intelligent sensors. In *Proceedings of 9th Symposium on Information Control in Manufacturing (INCOM'98)*, 24–26 June, 1998, Nancy-Metz, France, vol. III, 33–38.

[137] Gutkin L.S., *Electronic Devices Optimization by Set of Characteristics*, Sovetskoe Radio, Moscow, 1975 (in Russian).

[138] Yurish S.Y., Kirianaki N.V., Hardware/Software co-synthesis methodology for single-chip microcomputer based converters. In *Proceeding of International Conference Programmable Devices and Systems (PDS'95)*, 9–10 November 1995, Gliwice, Poland, 153–157.

[139] Altshyller G.B., Eltimov N.N., Shakylin V.G. *Economical-to-operate Mini Quartz-crystal Oscillators*, Sviaz, Moscow, 1979 (in Russian).

[140] Yurish S.Y., Kirianaki N.V., Shpak N.O., Reference's accuracy in virtual frequency-to-code converters. In *Proceedings of 5th International Workshop on ADC Modelling and Testing (IWADC'2000)*, parallel event to *XVI IMEKO World Congress*, Wien, Austria, 26–28 September, 2000, Vol.X, 325–330.

[141] Lazarev V.G., Piyl E.I., *Control Automata Synthesis*, Energiya, Moscow, 1970 (in Russian).

[142] Sovetov B.Y., Kutuzov O.L., Golovin Y.A., Avetov Y.V., *Microprocessors in Data-Transmission Systems*, Vyssaya Shkola, Moscow, 1987 (in Russian).

[143] Yurish S.Y., Kirianaki N.V., A modified method of algorithm merging for DSP. In *Proceedings of Workshop on Design Methodologies for Signal Processing*, 29–30 August, 1996, Zakopane, Poland, 81–84.

[144] Data Transmission Design Seminar, *Reference Manual*, Texas Instruments, 1998.

[145] Reiss L. *Introduction to Local Area Network with Microcomputer Experiments*, Prentice-Hall, Englewood Cliffs, 1987.

[146] Linsmeier, Klaus-Dieter *Sensor Systems for the Automobile*, Verlag Moderne Industrie, 1999.

[147] Car systems integrate more functions, *Electronic Components*, June, 1997, 186–226.

[148] Yurish S.Y., Kirianaki N.V., Shpak N.O. Novel rotation speed measurement concept for ABS appropriated for microsystem creation. In *Advanced Microsystems for Automotive Applications 99*, Edited by Detlef E. Ricken and Wolfgang Gessner, Springer, 1999, 215–223.

[149] Automotive sensors, *European Sensor News and Technology*, September, 1998, 11–12.

[150] *Automotive sensors*, Sensors Series, edited by M. H. Westrook and J.D. Turner, Institute of Physics Publishing, Bristol, UK, 1994.

[151] Kirianaki N.V., Yurish S.Y., Intelligent adaptive systems of telemetry. In *Proceedings of International Conference on Programmable Devices and Systems*, 26–28 November, 1996, Ostrava, Czech Republic, 224–231.

[152] Barzhin V.Y., Kolpakov F.F., Laskin O.V., Shmaliy Y.S., Digital Device for Information Processing of Two-Parametrics Frequency Sensors, *Patent 851279*, USSR, 1981.

[153] http://www.seba.de/html/mps-d_e.htm.

[154] Tompkins W.J., Webster J.G., *Interfacing Sensors to the IBM PC*, Prentice Hall, Englewood Cliffs, USA, 1988.

[155] Brignell J.E., Taner A.H., Virtual instrumentation for intelligent sensors: Development, production and in the field, *Jubilee Issue of Research Journal*, University of Southampton, (1997–1998) CD-ROM.

[156] *The Measurement and Automation Catalog*, 2001, National Instruments, USA.

[157] Winiecki W., Virtual instruments–What does it really mean? In *Proceedings of XIV IMEKO World Congress*, Topic 4 on the CD-ROM, 1–6 June, 1997, Tampere, Finland.

[158] Des-Jardin Larry, Virtual instruments and the role of software. In *Electronic Instrument Handbook*, McGraw-Hill, New York, 1993.

[159] Sobolev V. The new data acquisition technology on the basis of virtual instrumentation. In *Proceedings of XIV IMEKO World Congress*, Vol.5, Topic 7 on the CD-ROM, 1–6 June, 1997, Tampere, Finland.

[160] Kirianaki N.V., Shpak N.O., Yurish S.Y., Virtual instruments: From virtual interface toward virtual measuring channel. In *Proceedings of IEEE International Workshop on Emerging Technologies, Intelligent Measurements and Virtual System for Instrumentation and Measurements (ETIMVIS'98)*, 15–16 May, 1998, St Paul, USA, 133–137.

[161] *BURR-BROWN IC Data Book*, Data Conversion Product, 1995.

[162] Taubert P., *Evaluation of Accuracy of Results of Measurement*, Energoatomizdat, Moscow, 1988 (in Russian).

[163] Kirianaki N.V., Klushin Y.S., Computer method of construction of probability distribution of an error of two and more components, *Measuring Technique*, No. 1, 1986, 3–4 (in Russian).

[164] Kirianaki N.V., Shpak N.O., Yurish S.Y., Khoma V.V., Dzoba Y.S., New method of summation for measurement errors based on piece-wise linear approximation of probability distribution, *Journal of Electrical Engineering*, 51(3–4), 2000, 94–99.

[165] Krentz M., Carro L., Suzim A., System integration with dedicated processor for industrial applications. In *Proceedings of 3rd IFAC Symposium on Intelligent Components and Instruments for Control Applications*, Annecy, France, 9–11 June, 1997, 333–337.

[166] Kirianaki N.V., Shpak N.O., Yurish S.Y. Microcontroller cores for metering applications. In *Proceedings of International Conference on Programmable Devices and Systems (PDS'98)*, 24–25 February, 1998, Gliwice, Poland, 345–352.

[167] Furber S.B. Embedded system design. In *Low Power Design in Deep Submicron Electronics*, edited by W. Nebel and J. Mermet, Kluwer Academic Publishers, NATO ASI Series, Vol. 337, London, 1997, 397–417.

[168] Piguet C. Microprocessors design. In *Low Power Design in Deep Submicron Electronics*, edited by W. Nebel and J. Mermet, Kluwer Academic Publishers, NATO ASI Series, Vol. 337, London, 1997, 513–541.

[169] Koval V.A., Shevchenko V.O., Shpak N.O., Yurish S.Y., Low power design technique for embedded microcontroller cores. In *Proceedings of 4th International Workshop Mixed Design of Integrated Circuits and Systems (MIXDES' 97)*, 12–14 June, 1997, Poznan, Poland, 619–624.

[170] Roy K., Jonson M.C., Software design for low power. In *Low Power Design in Deep Submicron Electronics*, edited by W. Nebel and J. Mermet, Kluwer Academic Publishers, NATO ASI Series, Vol. 337, London, 1997, 433–460.

[171] Sato T., Ootaguo Y., Nagamatsu M., Tago H. Evaluation of architecture-level power estimation for CMOS RISC processors. In *Proceedings of the Symposium on Low Power Electronics*, 1995, 44–45.

[172] Su C.-L., Tsui C.-Y., Despain A.M., Low power architecture design and compilation techniques for high-performance processors. In *Proceedings of IEEE COMPCON*, 1994, 489–498.

[173] Tiwari V., et al., Power analysis of embedded software: A first step towards software power minimization, *IEEE Transaction on VLSI Systems*, 2(4), December, 1994, 437–445.

[174] Lee M. T.-C., Tiwari V., Malik S., Fujita M., Power analysis and minimization techniques for embedded DSP software, *Fujitsu Scientific and Technical Journal*, 31(2), December, 1995, 215–229.

[175] Manne S., Pardo A., Bahar R.I., Hachtel G.D., Somenzi F., Macii E., Poncino M., Computing the maximum power cycles of a sequential circuit. In *Proceedings of the 32nd Design Automation Conference*, June, 1995, 23–28.

[176] Meijer G., Herwaarden S., Jong G., Toth F., Li X. Intelligent Sensor System and Smart Sensor: Applications, In *Book Low-Power HF Microelectronics. A unified approach*, Ed. by Gerson A.S. Machado, IEEE, 1996, 977–1018.

[177] Nazarov S. V., Barsukov A. G. *Measuring Devices and Optimization of Computer Systems*, Moscow, Radio i svyaz', 1990 (in Russian).

[178] Huijsing J.H., Maijer G.C.M., *Smart Sensor Interfaces*, Kluwer Academic Publishers, 1997.

[179] MCM20027/D, *Motorola Semiconductor Technical Data*, 2000.

[180] Sensors. Control, *Standard Products Catalog*, Trafag GmbH, 2001.

[181] http://www.madur.com

[182] *CAN Newsletter*, No.1, March, 2001, p.38.

[183] *The 2000–2001 CAN Solution Directory*, 2000.

[184] van der Goes, Frank M.L. Meijer, Gerard C.M. A universal transducer interface for capacitive and resistive sensor elements, *Analog Integrated Circuits and Signal Processing*, 14, 1997, 249–260.

[185] http://www.smartec.nl.

[186] Universal Transducer Interface (UTI), *Application Note of UTI*, Smartec, 1998.

[187] Universal Transducer Interface (UTI), *Revolution in Sensor Interfacing*, Smartec, 1998.

[188] *Acam –Solutions in Time, Precision Time Interval Measurement*, 2000.

Appendix A

WHAT IS ON THE SENSORS WEB PORTAL?

The monthly up-dated Sensors Web Portal is located at www.sensorsportal.com. The site aims to be a central source of information on smart sensors and includes the following sections:

- Sensors section including temperature, pressure, flow, humidity, chemical, gas, optical, rotation speed, magnetic, mechanical and other subsections
- archives with full-page articles, references and abstracts
- standardization
- patents
- the marketplace
- projects
- training courses
- tools
- links
- *Sensors & Transducers e-Digest.*
- Bookstore
- Wish List

It is a snapshot of what is happening in the field of smart sensors. Information is edited from the best international printed and electronic media, using research, application, novelty and innovation as key words. We hope that you will find the sensors web portal interesting and that strong links will be established among our members involved in sensors R&D and production.

GLOSSARY OF SMART SENSOR TERMS

Absolute pressure sensor — A sensor that measures the input pressure in relation to zero pressure (a total vacuum on one side of the diaphragm).

Accuracy — (1) The degree of conformity of a measured or calculated value to its definition or with respect to a standard reference (*see* uncertainty). (2) The maximum error of a measured value with respect to its true theoretical value.

Acceleration — The rate of change of velocity. Acceleration has two characteristics: magnitude and direction.

Acquisition time — The time required for the front end of a DAQ board to capture an input signal and hold it to within a specified error band after a sample command is received.

A/D — Analog-to-digital

ADC — Analog-to-digital converter–An electronic device, often an integrated circuit that converts an analog voltage to a digital number.

Analog-to-digital (A/D) conversion — The process of converting a continuous analog signal to a digital value that represents that signal at the instant at which it was sampled.

Algorithm — A well-defined procedure that transforms one or more given input variables into one or more output variables in finite numbers of steps.

ASIC — Application-Specific Integrated Circuit–A proprietary semiconductor component designed and manufactured to perform a set of specific functions for a specific customer.

Asynchronous — (1) Hardware–A property of an event that occurs at an arbitrary time, without synchronization to a reference clock. (2) Software–A property of a function that begins an operation and returns prior to the completion or termination of the operation.

Asynchronous communications — Communications protocols in which data are transmitted at arbitrary points in time. Most commonly used to refer to serial data transmission protocols such as RS-232 or RS-485.

Background acquisition — Data is acquired by a DAQ system while another program or processing routine is running without apparent interruption.

Biosensor — A sensor that either detects a biological substance or incorporates biological materials to accomplish sensing.

Calibration — A test during which known values of a measurand are applied to the device under test and corresponding output readings are recorded under specified conditions.

CAN — Controller Area Network–A serial bus that finds increasing use as a device-level network for industrial automation. CAN was developed by Bosch to address the needs of in-vehicle automotive communications.

Compensation — The technique of modifying data from a source to correct the influence of additional environmental effects.

Conversion rate — The speed of a data acquisition system expressed in a number of conversions or samples per second.

Conversion time — The time required, in an analog input or output system, from the moment a channel is interrogated (like in a read instruction) to the moment that accurate data are available.

Counter/Timer — A circuit that counts external pulses or clock pulses (timing) and can either operate as an event counter or measure the time between two events.

D/A — Digital-to-Analog

DAC — Digital-to-analog converter–An electronic device, often an integrated circuit that converts a digital number into a corresponding analog voltage or current.

DAQ — Data acquisition–(1) Collecting and measuring electrical signals from sensors, transducers, and test probes or fixtures and inputting them into a computer for processing; (2) Collecting and measuring the same kinds of electrical signals with A/D and/or DIO boards plugged into a PC, and possibly generating control signals with D/A and/or DIO boards in the same PC.

Data acquisition board — A data acquisition system incorporated on a PCB that is electrically and mechanically compatible with a particular computer system.

Data acquisition system — A system that processes one or more analog or quasi-digital signals and converts them into a digital form for the use by a computer system.

Data logger — A data acquisition system that incorporates a small computer, is typically portable, and is intended to collect data autonomously for extended periods of time. The data are afterwards downloaded into another computer for processing and analysis.

Differential pressure sensor — A sensor, which is designed to accept simultaneously two independent pressure sources. The output is proportional to the pressure difference between the two sources.

Digital-to-analog (D/A) conversion — The process of converting a digital signal or a code into an analog or quasi-digital signal.

Digital-to-analog converter (DAC) — A device that converts a digital value or code into an analog or quasi-digital signal.

Digital output — Output that is of only two stable states, appearing in the manner of a switch; that is, it is either On or Off or High or Low (i.e., high voltage or low voltage).

Discrete Fourier transform — A version of the Fourier transform that operates on data that have been sampled at discrete, uniformly spaced points in time.

DMA — Direct Memory Access–A method by which data can be transferred to/from a computer memory from/to a device or memory on the bus while the processor does something else. A DMA is the fastest method of transferring data to/from a computer memory.

Drift (frequency) — The linear (first-order) component of a systematic change in frequency of an oscillator over time. Drift occurs due to ageing plus changes in the environment and other factors external to the oscillator.

DSP — (Digital signal processing)–(1) the science concerned with representation of signals by sequences of numbers and the subsequent processing of these number sequences. (2) Techniques for modifying and analysing a signal after it has been sampled and converted into the digital domain by an ADC.

Dynamic error — The error which occurs because the sensor's output does not precisely follow the variations in time of the measurand.

Dynamic range — The ratio of the largest signal a system can handle to the smallest signal it can reliably resolve. A dynamic range is typically expressed in decibels for analog systems and bits (N) for digital systems, where dB = 6.02 N = 20 log (the largest signal/the smallest signal resolved).

Encoder — A device that converts the linear or rotary displacement into digital or pulse signals. The most popular type of encoder is the optical encoder, which uses a rotating disk with alternating opaque areas, a light source and a photodetector.

Error — The difference of a measured value from its known true or correct value (or sometimes from its predicted value).

4-Wire resistance measurement — A way to measure the values of a resistor while avoiding errors caused by the wire runs. Two wires carry a current to the resistor, and two wires measure the voltage generated. Commonly used with resistance temperature detectors (RTDs).

Fourier transform — A mathematical technique that transforms a continuous function from its time-domain representation to its frequency-domain representation. The discrete Fourier transform performs the analogous function on discretely sampled data.

Frequency — The rate at which a periodic phenomenon occurs over time.

Frequency deviation — The difference between frequency values of the same signal at two different times or the difference between the instantaneous signal frequency and the average signal frequency.

Frequency difference — Difference between the frequencies of two different signals.

Frequency output — An output in the form of frequency, which varies as a function of the applied measurand.

Full scale (FS) — The maximum specified range of a data acquisition system.

Full-scale range (FSR) — The difference between minimum and maximum allowable input or output values for a data acquisition system.

General-Purpose Interface Bus (GPIB) — IEEE-488 standard interface connecting peripheral devices, often sensors and programmable instruments, to a computer.

GUI — graphical user interface–An intuitive, easy-to-use means of communicating information to and from a computer program by means of graphical screen displays. GUIs can resemble front panels of instruments or other objects associated with a computer program.

Hall effect — When a semiconductor, through which a current is flowing, is placed in a magnetic field, a difference in potential (voltage) is generated between the two opposed edges of the conductor in the direction mutually perpendicular to both the field and the conductor. Typically used in sensing magnetic fields.

Hysteresis — The measure of a sensor's ability to represent changes in the input parameter, regardless of whether the input is increasing or decreasing.

Intelligent sensor — *See smart sensor*

Integrating ADC — An ADC in which the input voltage is integrated over time. Different types of ADCs include a single slope, a duel slope, a quad slope, and a charge balancing.

Integrated circuit (IC) — An interconnected array of active and passive elements integrated within a single semiconductor substrate or other compatible material, and capable of performing one complete electronic function.

Interface — A common boundary between electronic systems, or parts of a single system.

Interface circuit — A circuit that links one type of a device with another. Its function is to produce the required current and voltage levels for the next stage of the circuitry from the previous stage.

Linearity (linearity error) — The deviation of the sensor output curve from a specified straight line. A linearity error is usually expressed as a percent of the full-scale output.

Linearization — The process of modifying a signal, either analog or digital, to compensate for the nonlinearities present in the source or previous signal processing.

Measurand — A physical quantity, property or condition, which is measured (e.g., pressure, acceleration).

Modulating Sensor — *See parametric sensor.*

MEMS — An IC chip that provides sensing and/or actuation functions in addition to electronic ones.

Noise — An output signal of the random amplitude and random frequency not present in the measurand.

Offset — The difference between the realized value and the reference value.

Parametric (modulating) sensor — A device producing the primary information by way of respective alterations of any electrical parameters of some electrical circuit (inductance, capacity, resistance, etc.), the measuring of which it is necessary to have an external auxiliary power supply. Examples of such types of sensors are pressure sensors based on the piezoresistive effect and photodetectors based on the photoelectric effect. Sometimes the modulating sensor is called the 'passive' sensor.

PCM (Program-oriented conversion method) — The processor algorithm of measurement, incarnated in the functional-logic structure of a computer or a microcontroller through the software.

Piezoelectric effect — The property of certain materials that allows them to develop a voltage when deformed by stress, or to become strained when subjected to the application of a voltage.

Piezoresistive effect — The property of a resistor that produces a change in resistance in response to the applied strain.

Precision — The degree of mutual agreement among a series of individual measurements. Precision is often, but not necessarily, expressed by the standard deviation of measurements.

Pressure sensor — A device that converts an input pressure into an electrical output.

Proximity sensor — A device that detects the presence of an object without physical contact. Most proximity sensors provide a digital on/off relay or a digital output signal.

PWM — Pulse-width modulation–Generation of a pulse waveform with a fixed frequency and variable pulse width (the duty-cycle). PWM is used to control discrete devices such as DC motors and heaters by varying the pulse width (the ratio of on time to off time).

Quantization error — The inherent uncertainty in digitizing an analog value due to the finite resolution of the conversion process.

Quasi-digital sensor — The discrete frequency-time domain sensor with the frequency, the period, the duty-cycle, the time interval, the pulse number or the phase shift output.

Range — The measurand values over which the sensor is intended to measure, specified by the upper and lower limits.

Real-time processing — A procedure in which results of an acquired and computed value can be used to control a related physical process in real time.

Reference — A stable source for a physical quantity, such as voltage, frequency, etc. used in a measuring device to maintain measurement stability and repeatability.

Relative accuracy — A measure in LSB of the accuracy of an ADC. It includes all nonlinearity and quantization errors. It does not include offset and gain errors of the circuitry feeding the ADC.

Reliability — The measure of a sensor's ability to maintain both accuracy and precision under conditions for which it is designed to perform for the expected life of the device.

Resolution — The smallest significant difference that can be measured with a given instrument. Resolution can be expressed in bits, in proportions, or in a percent of a full scale. For example, a system has a 12-bit resolution, one part in 4.096 resolution, and 0.0244 percent of a full scale; a measurement made with a time interval counter might have a resolution of 10 ns.

Response time — The time needed for a sensor to register a change (within a tolerance of an error) in the parameter it is measuring.

Self-calibration — A property of a sensor that has an extremely stable reference and calibrates its own ADC without manual adjustments by the user.

Self-generating sensor — The device permitting to receive a signal immediately by the way of a current $i(t)$ or voltage $V(t)$ and which does not require any source of power other than the signal being measured. Examples of such types of sensors are Seebeck-effect based thermocouples and photo-effect based solar cells. Self-generating sensors are also called 'active' sensors.

Sensitivity — The minimum change in the parameter being measured that will produce a detectable change in a sensor's output.

Sensor — The basic element that usually changes some physical parameter (heat, light, sound, pressure, motion, flow, etc.) to a corresponding electrical signal.

Sensing element — That part of a sensor which responds directly to changes in the input pressure.

Signal conditioning — The processing of the form or mode of a signal so as to make it intelligible to or compatible with a given device, including such manipulation as pulse shaping, pulse clipping, digitizing and linearizing.

Smart sensor — One chip, without external components, including the sensing, interfacing, signal processing and intelligence (self-testing, self-identification or self-adaptation) functions.

SS — Simultaneous Sampling–A property of a system in which each input or output channel is digitized or updated at the same instant.

Strain guage — A piezoresistive sensing device providing a change in the electrical resistance proportional to the level of the applied stress.

Synchronous — (1) Hardware–A property of an event that is synchronized to a reference clock. (2) Software–A property of a function that begins an operation and returns only when the operation is complete.

Thermistor — A device that measures temperature-induced changes in resistance of a resistor or a semiconductor.

Thermocouple — A temperature-measuring device made of two dissimilar conductors joined together at their ends. The unit generates the thermoelectric voltage between the junctions that represents their temperature difference.

Telemetry — Transmission–via radio waves, wires, etc.–of the instrument reading across distances. Also called telemetering or remote metering.

Transducer — A fully packaged, signal-conditioned, compensated and calibrated sensor.

Transfer function — The input-to-output response characteristics of a device.

Transmitter — A device that converts the output of a sensor into a form more suitable for communication to another system.

Uncertainty — Limits of the confidence interval of a measured or calculated quantity. Note: The probability of the confidence limits should be specified, preferably as one standard deviation.

Virtual instrument — A measuring instrument composed of a general-purpose computer equipped with cost-effective measurement hardware blocks (internal and/or external) and software, that performs functions of a traditional instrument determined both by the hardware and the software, and operated by means of specialized graphics on a computer screen. The necessary condition of a virtual instrument existing is the software realization of the user interface, performed by a general-purpose computer and the sufficient condition is that a hardware and a software part of the virtual instrument do not exist separately as an instrument.

Voltage-to-frequency converter (VFC) — A device that converts an input voltage into a periodic waveform output with the frequency proportional to the input voltage.

Wiegand-effect sensor — The generation of an electrical pulse in a coil wrapped around or located near a Wiegand (a specially processed ferromagnetic) wire that is subjected to a changing magnetic field. The effect is proprietary and patented.

INDEX

Printed and bound in the UK by
CPI Antony Rowe, Eastbourne

Printed and bound by CPI Group (UK) Ltd, Croydon, CR0 4YY

27/10/2024

14580218-0002